Exercise Solutions
to Accompany

BASIC
CIRCUIT
ANALYSIS

SECOND EDITION

David R. Cunningham
John A. Stuller
University of Missouri-Rolla

Prepared By:
Tonda Davis
Frieda Adams
Dr. Khanh T. Ngo
University of Missouri-Rolla

JOHN WILEY & SONS, INC.

NEW YORK • CHICHESTER • BRISBANE • TORONTO • SINGAPORE

Acknowledgments

Special thanks are due to Dr. Khanh T. Ngo for checking the accuracy of the exercise solutions. The authors also thank Tonda Davis for all computer graphics and typing the manuscript.

Table of Contents for Exercise Solutions

Part One   **Fundamentals**

Chapter 1   Introduction   1

Chapter 2   Network Laws and Components   4

Chapter 3   Introduction to Circuit Analysis   19

Part Two   **Time Domain Circuit Analysis**

Chapter 4   Node Voltages and Mesh Currents   31

Chapter 5   Network Properties   47

Chapter 6   Operational Amplifiers   58

Chapter 7   Signal Models   65

Chapter 8   First-Order Circuits   70

Chapter 9   Second-Order Circuits   84

Part Three   **Frequency Domain Analysis**

Chapter 10   The Sinusoidal Steady State   94

Chapter 11   AC Circuit Analysis   100

Chapter 12   Power in AC Circuits   109

Chapter 13   Frequency Response   113

Part Four   **Transform Methods**

Chapter 14   The Laplace Transform   132

Chapter 15   Fourier Series   144

Part Five   **Selected Topics**

Chapter 16   Equivalent Circuits for Three-Terminal Networks
               and Two-Port Networks   161

Chapter 17   Mutual Inductance and Transformers   174

Chapter 18   Single- and Three-Phase Power Circuits   182

CHAPTER 1

Introduction

Exercises

1. (a) $0.002 \text{ m} = 2 \times 10^{-3} \text{ m} = 2 \text{ mm}$

    (b) $15,000 \text{ m} = 15 \times 10^3 \text{ m} = 15 \text{ km}$

    (c) $0.05 \text{ s} = 50 \times 10^{-3} \text{ s} = 50 \text{ ms}$

    (d) $0.0004 \text{ ms} = 400 \times 10^{-9} \text{ s} = 400 \text{ ns}$

    (e) $27,000 \text{ kW} = 27 \times 10^6 \text{ W} = 27 \text{ MW}$

    (f) $0.007 \text{ s} = 7 \times 10^{-3} \text{ s} = 7 \text{ ms}$

2. (a) $p = 120\sqrt{2}\cos 2\pi 60t$ is time varying so a lower case p is used. ***Consistent***

    (b) $I = 5t^2$ is time varying so a lower case i should be used. ***Inconsistent***

    (c) $p = 12$ is a constant, but if this represents instantaneous power rather than average power, a lower case p should be used because instantaneous power can be time varying. ***Consistent***

    (d) $I = 12\sin 2t$ is time varying so a lower case i should be used. ***Inconsistent***

3. $q = \int_0^5 i\,dt = \int_0^5 200\,dt = 1000 \text{ C} = 1 \text{ kC}$

4.   (a) $i_{12} = 10e^{-2t}$ A $q = \int_0^t i\,d\lambda = \int_0^t 10e^{-2\tau}\,d\lambda = 5(1 - e^{-2t})$ C

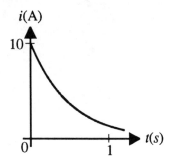

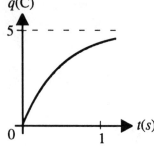

(b) $i = 10\cos 2\pi t$ A $q = \int_0^t i\,d\lambda = \int_0^t 10\cos 2\pi\lambda\,d\lambda = \dfrac{10}{2\pi}\sin 2\pi t$ C

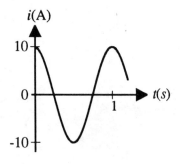

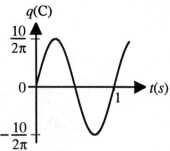

5.   (a) $q = 10e^{-2t}$ C $i = \dfrac{d}{dt}q = \dfrac{d}{dt}10e^{-2t} = -20e^{-2t}$ A

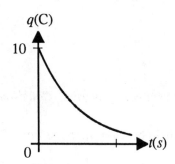

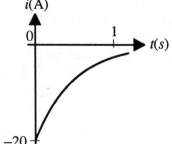

(b) $q = 10\cos 2\pi t$ C $i = \dfrac{d}{dt}q = \dfrac{d}{dt}10\cos 2\pi t = -20\pi\sin 2\pi t$ A

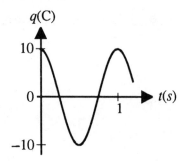

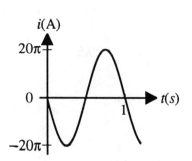

6. $v = \dfrac{50\ J}{4\ C} = 12.5\,\dfrac{J}{C} \;\Rightarrow\; \boxed{v = 12.5\ \text{V}}$

7. $w = (10 \text{ V})(1,000 \text{ C}) = \left(10\frac{\text{J}}{\text{C}}\right)(1,000 \text{ C}) = 10,000 \text{ J} \Rightarrow \boxed{w = 10 \text{ kJ}}$

8. (a) Arrow enters (-) and exits (+) mark. *Active*

   (b) Arrow enters (+) and exits (-) mark. *Passive*

   (c) Arrow enters (-) and exits (+) mark. *Active*

   (d) Arrow enters (+) and exits (-) mark. *Passive*

   (e) Arrow enters (+) and exits (-) mark. *Passive*

9. (a) $i_1 = -\dfrac{p_1}{v_1} = -\dfrac{1}{-12} \Rightarrow \boxed{i_1 = -\dfrac{1}{12} \text{ A}}$

   (b) $i_2 = \dfrac{p_2}{v_2} = \dfrac{2}{\frac{24}{5}} \Rightarrow \boxed{i_2 = \dfrac{5}{12} \text{ A}}$

   (c) $v_3 = -\dfrac{p_3}{i_3} = -\dfrac{3}{-\frac{5}{12}} \Rightarrow \boxed{v_3 = \dfrac{36}{5} \text{ V}}$

   (d) $i_4 = \dfrac{p_4}{v_4} = \dfrac{4}{12} \Rightarrow \boxed{i_4 = \dfrac{1}{3} \text{ A}}$

   (e) $i_B = \dfrac{p_B}{v_B} = \dfrac{-8}{12} \Rightarrow \boxed{i_B = -\dfrac{2}{3} \text{ A}}$

10. (a) $p_A = -(48)(2) \Rightarrow \boxed{p_A = -96 \text{ W}}$

    (b) $p_B = (12)(-2) \Rightarrow \boxed{p_B = -24 \text{ W}}$

    (c) $p_C = -(-60)(8) \Rightarrow \boxed{p_C = 480 \text{ W}}$

    (d) $p_D = (80)(-6) \Rightarrow \boxed{p_D = -480 \text{ W}}$

    (e) $p_E = (20)(6) \Rightarrow \boxed{p_E = 120 \text{ W}}$

    (f) $p_A + p_B + p_C + p_D + p_E = (-96) + (-24) + (480) + (-480) + (120) = 0$

11. $p = vi = (10)(200) = 2000 \text{ W} \Rightarrow \boxed{p = 2 \text{ kW}}$

12. $P = \dfrac{1}{t_2 - t_1}\int_{t_1}^{t_2} p\,dt = \dfrac{1}{20}\left[\int_0^5 (200)10\,dt + \int_5^{10} 0\,dt + \int_{10}^{20}(200)10\,dt\right]$

    $= \dfrac{1}{20}[10,000 + 0 + 20,000] = 1,500 \text{ W} \Rightarrow \boxed{P = 1.5 \text{ kW}}$

CHAPTER 2

Network Laws and Components

Exercises

1.   (a)   There are 5 nodes (0, 1, 2, 3, 4)
               There are 4 junctions (0, 1, 2, 3)

     (b) A and B are in parallel
               C and D are in parallel
               G and I are in parallel

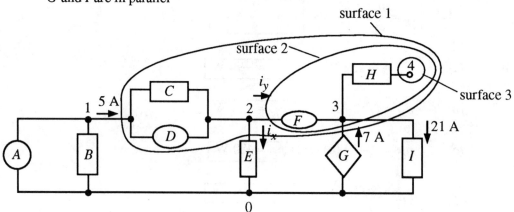

     (c)   KCL for closed surface 1

$$-5 + i_x - 7 + 21 = 0 \implies \boxed{i_x = -9\text{A}}$$

KCL for closed surface 2

$$-i_y - 7 + 21 = 0 \implies \boxed{i_y = 14\text{A}}$$

     (d) $i_{32} = -i_y = -14\text{A}$

KCL for closed surface 3

$$-i_{34} = 0 \implies \boxed{i_{34} = 0}$$

2.

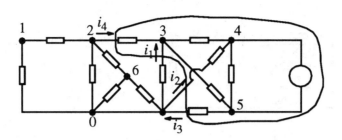

(a) There are 7 nodes (0, 1, 2, 3, 4, 5, 6)
   There are 6 junctions (0, 2, 3, 4, 5, 6)

(b) The two left-hand components are in series.
   The two right-hand components are in parallel.

(c) For the indicated closed surface

$$-i_4 - i_1 - i_2 + i_3 = 0$$

$$i_4 = -i_1 - i_2 + i_3 = -3 - (-4) + 5 \implies \boxed{i_4 = 6 \text{ A}}$$

3.

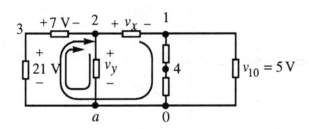

(a) There are 5 nodes (0, 1, 2, 3, 4)
   There are 3 junctions (0, 1, 2)

(b) The two left-hand components are in series.
   The components from node 1 to 4 and 4 to 0 are in series.

(c) There are 3 meshes. (a, 3, 2,), (a, 2, 1, 4, 0), (0,4,1)

(d) There are 6 loops.

(e) KVL gives

$$v_x + v_{10} - 21 + 7 = 0$$

$$v_x = -5 + 21 - 7 \implies \boxed{v_x = 9 \text{ V}}$$

   KVL also gives

$$v_y - 21 + 7 = 0$$

$$v_y = 21 - 7 \implies \boxed{v_y = 14 \text{ V}}$$

4.

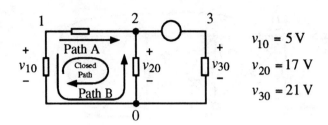

$$v_{10} = 5 \, V$$
$$v_{20} = 17 \, V$$
$$v_{30} = 21 \, V$$

(a) $\quad -v_{10} + v_{12} + v_{20} = 0$

$$v_{12} = v_{10} - v_{20} = 5 - 17 \implies \boxed{v_{12} = -12 \, V}$$

(b) $\quad v_{12} = v_{10} - v_{20} = 5 - 17 \implies \boxed{v_{12} = -12 \, V}$

(c) $\quad v_{13} = v_{10} - v_{30} = 5 - 21 \implies \boxed{v_{13} = -16 \, V}$

5.

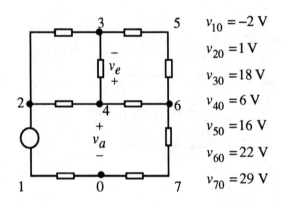

$$v_{10} = -2 \, V$$
$$v_{20} = 1 \, V$$
$$v_{30} = 18 \, V$$
$$v_{40} = 6 \, V$$
$$v_{50} = 16 \, V$$
$$v_{60} = 22 \, V$$
$$v_{70} = 29 \, V$$

Use KVL

$$v_a = v_{40} \implies \boxed{v_a = 6 \, V}$$
$$v_{21} = v_{20} - v_{10} = 1 - (-2) \implies \boxed{v_{21} = 3 \, V}$$
$$v_{12} = -v_{21} \implies \boxed{v_{12} = -3 \, V}$$
$$v_{53} = v_{50} - v_{30} = 16 - 18 \implies \boxed{v_{53} = -2 \, V}$$

$$v_{16} = v_{10} - v_{60} = -2(-22) \implies \boxed{v_{16} = -24 \, V}$$
$$v_{67} = v_{60} - v_{70} = 22 - 29 \implies \boxed{v_{67} = -7 \, V}$$
$$v_e = v_{40} - v_{30} = 6 - 18 \implies \boxed{v_e = -12 \, V}$$
$$v_{14} = v_{10} - v_{40} = -2 - 6 \implies \boxed{v_{14} = -8 \, V}$$

6.

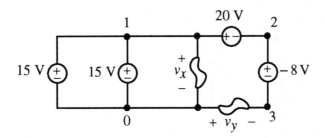

The sources and KVL give

$$\boxed{v_{10} = 15 \text{ V}}$$

$$v_x = v_{10} \Rightarrow \boxed{v_x = 15 \text{ V}}$$

$$v_{13} = 20 + (-8) \Rightarrow \boxed{v_{13} = 12 \text{ V}}$$

$$v_{20} = -20 + 15 \Rightarrow \boxed{v_{20} = -5 \text{ V}}$$

$$v_y = -15 + 20 - 8 \Rightarrow \boxed{v_y = -3 \text{ V}}$$

7.

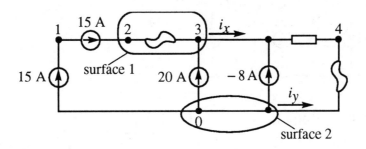

The sources and KCL give

$$\boxed{i_{10} = -15 \text{ A}}$$

$$\boxed{i_{23} = 15 \text{ A}}$$

KCL for closed surface 1:

$$-15 - 20 + i_x = 0 \Rightarrow \boxed{i_x = 35 \text{ A}}$$

KCL for closed surface 2 gives

$$15 + 20 + (-8) + i_y = 0 \Rightarrow \boxed{i_y = -27 \text{ A}}$$

8. (a) Given the circuit below:

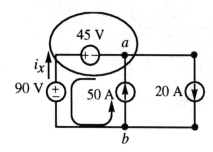

From KCL: $i_x = -50 + 20 = 0 \Rightarrow \boxed{i_x = -30 \text{ A}}$

From KVL: $v_{ab} = -45 + 90 \Rightarrow \boxed{v_{ab} = 45 \text{ V}}$

$p = (20)(v_{ab}) \Rightarrow \boxed{p = 900 \text{ W}}$

(b) Given the circuit below:

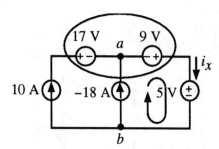

From KCL: $-10 + 18 + i_x = 0 \Rightarrow \boxed{i_x = -8 \text{ A}}$

From KVL: $v_{ab} = -9 + 5 \Rightarrow \boxed{v_{ab} = -4 \text{ V}}$

$p = (5)(i_x) \Rightarrow \boxed{p = -40 \text{ W}}$

(c) Given the circuit below:

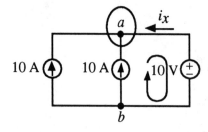

From KCL: $-10 - 10 - i_x = 0 \Rightarrow \boxed{i_x = -20 \text{ A}}$

From KVL: $\boxed{v_{ab} = 10 \text{ V}}$

$p = -(10)(i_x) \Rightarrow \boxed{p = 200 \text{ W}}$

(d) Given the circuit below:

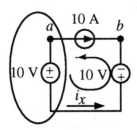

From KCL: $10 + i_x = 0 \Rightarrow \boxed{i_x = -10 \text{ A}}$

From KVL: $v_{ab} = 10 + 10 \Rightarrow \boxed{v_{ab} = 20 \text{ V}}$

$p = (10)(i_x) \Rightarrow \boxed{p = -100 \text{ W}}$

9.　　(a)　　Given the circuit below:

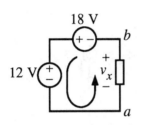 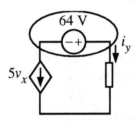

From KVL: $v_{ab} = -v_x = -12 + 18 \Rightarrow \boxed{v_{ab} = 6 \text{ V}}$

From KCL: $i_y + 5v_x = 0 \Rightarrow \boxed{i_y = 30 \text{ A}}$

(b)　　Given the circuit below:

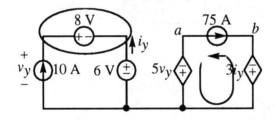

From KVL: $v_y = 8 + 6 = 14 \text{ V}$

From KCL: $-10 - i_y = 0 \Rightarrow \boxed{i_y = -10 \text{ A}}$

From KVL: $v_{ab} = -5v_y + 3i_y \Rightarrow \boxed{v_{ab} = -100 \text{ V}}$

(c)   Given the circuit below:

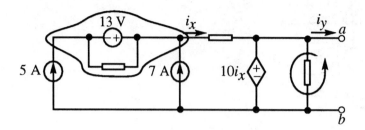

From KCL: $-5 - 7 + i_x = 0 \Rightarrow i_x = 12$ A

From KVL: $v_{ab} = 10(i_x) \Rightarrow \boxed{v_{ab} = 120 \text{ V}}$

From KCL: $\boxed{i_y = 0}$ since terminals a and b are open circuited

(d)   Given the circuit below:

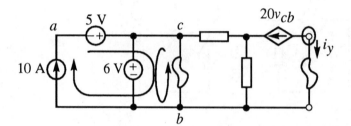

From KVL: $v_{ab} = -5 + 6 \Rightarrow \boxed{v_{ab} = 1 \text{ V}}$

From KVL: $v_{cb} = 6$ V

From KCL: $i_y + 20v_{cb} = 0 \Rightarrow \boxed{i_y = -120 \text{ A}}$

10.

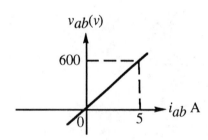

$R = \dfrac{\Delta v}{\Delta i} = \dfrac{600}{5} \Rightarrow \boxed{R = 120 \ \Omega}$

11.   Consider the volt-ampere measurement shown below.

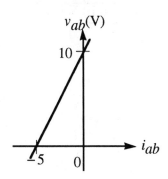

The slope of the line is constant, so the terminal equation is of the form:

$$v_{ab} = v_{oc} + R_{ab}i_{ab},$$

where $v_{oc} = v_{ab}|_{i_{ab}=0} \Rightarrow \boxed{v_{oc} = 10 \text{ V}}$

$$R_{ab} = \frac{\Delta v}{\Delta i} = \frac{10}{5} \Rightarrow \boxed{R_{ab} = 2 \ \Omega} \Rightarrow$$

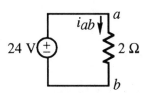

12.   (a)   Consider the circuit shown below:

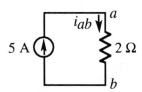

From KVL: $\boxed{v_{ab} = 24 \text{ V}}$

From Ohm's law: $i_{ab} = \dfrac{v_{ab}}{R} = \dfrac{24}{2} \Rightarrow \boxed{i_{ab} = 12 \text{ A}}$

$p = v_{ab}i_{ab} = (24)(12) \Rightarrow \boxed{p = 288 \text{ W}}$

(b)   Consider the circuit shown below:

From KCL: $\boxed{i_{ab} = 5 \text{ A}}$

From Ohm's law: $v_{ab} = Ri_{ab} = (2)(5) \Rightarrow \boxed{v_{ab} = 10 \text{ V}}$

$p = v_{ab}i_{ab} = (10)(5) \Rightarrow \boxed{p = 50 \text{ W}}$

(c)   Given the circuit shown below:

18 V i_{ab} a   3 S   b

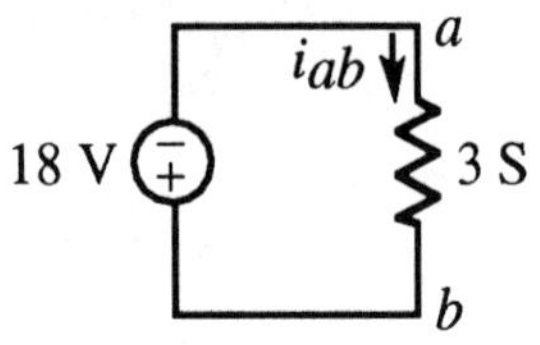

From KVL: $\boxed{v_{ab} = -18 \text{ V}}$

From Ohm's law: $i_{ab} = Gv_{ab} = (3)(-18) \Rightarrow \boxed{i_{ab} = -54 \text{ A}}$

$p = v_{ab}i_{ab} = (-18)(-54) \Rightarrow \boxed{p = 972 \text{ W}}$

(d)   Given the circuit shown below:

31 V   6 Ω   a   b   + −11 V −

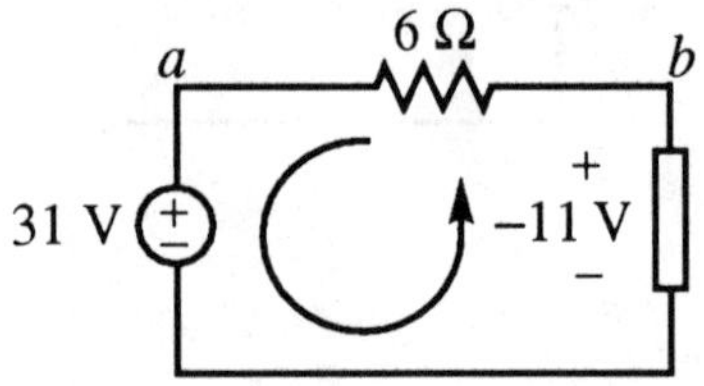

From KVL: $v_{ab} = 31 - (-11) \Rightarrow \boxed{v_{ab} = 42 \text{ V}}$

From Ohm's law: $i_{ab} = \dfrac{v_{ab}}{R} = \dfrac{42}{6} \Rightarrow \boxed{i_{ab} = 7 \text{ A}}$

$p = v_{ab}i_{ab} = (42)(7) \Rightarrow \boxed{p = 294 \text{ W}}$

(e)   Consider the following circuit.

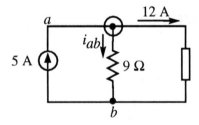

From KCL: $-5 + i_{ab} + 12 = 0 \Rightarrow \boxed{i_{ab} = -7 \text{ A}}$

From Ohm's law: $v_{ab} = Ri_{ab} = (9)(-7) \Rightarrow \boxed{v_{ab} = -63 \text{ V}}$

$$p = v_{ab}i_{ab} = (-63)(-7) \Rightarrow \boxed{p = 441 \text{ W}}$$

(f)   Given the circuit below:

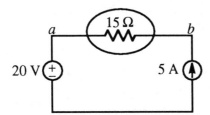

From KCL: $\boxed{i_{ab} = -5 \text{ A}}$

From Ohm's law: $v_{ab} = Ri_{ab} = (15)(-5) \Rightarrow \boxed{v_{ab} = -75 \text{ V}}$

$$p = v_{ab}i_{ab} = (-75)(-5) \Rightarrow \boxed{p = 375 \text{ W}}$$

(g)   Given the circuit below:

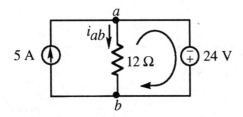

From KVL: $\boxed{v_{ab} = -24 \text{ V}}$

From Ohm's law: $i_{ab} = \dfrac{v_{ab}}{R} = \dfrac{-24}{12} \Rightarrow \boxed{i_{ab} = -2 \text{ A}}$

$$p = v_{ab}i_{ab} = (-24)(-2) \Rightarrow \boxed{p = 48 \text{ W}}$$

(h)   Given the circuit shown below:

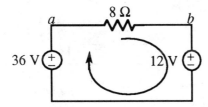

From KVL: $v_{ab} = 36 - 12 \Rightarrow \boxed{v_{ab} = 24 \text{ V}}$

From Ohm's law: $i_{ab} = \dfrac{v_{ab}}{R} = \dfrac{24}{8} \Rightarrow \boxed{i_{ab} = 3 \text{ A}}$

$$p = v_{ab}i_{ab} = (24)(3) \Rightarrow \boxed{p = 72 \text{ W}}$$

(i)   Given the circuit shown below:

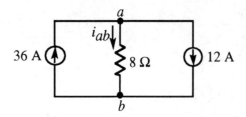

From KCL: $-36 + i_{ab} + 12 = 0 \Rightarrow \boxed{i_{ab} = 24 \text{ A}}$

From Ohm's law: $v_{ab} = Ri_{ab} = (8)(24) \Rightarrow \boxed{v_{ab} = 192 \text{ V}}$

$p = v_{ab}i_{ab} = (192)(24) \Rightarrow \boxed{p = 4608 \text{ W}}$

13.   Determine i_2 in the following circuit.

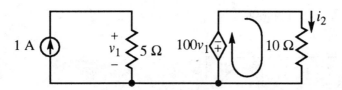

From Ohm's law: $v_1 = (5)(1) = 5 \text{ V}$

From KVL: $100v_1 + 10i_2 = 0 \Rightarrow i_2 = \dfrac{-100v_1}{10} = \dfrac{-500}{10} \Rightarrow \boxed{i_2 = -50 \text{ A}}$

14.   Determine the voltage v_2 in the circuit shown below.

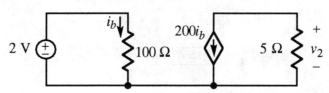

From Ohm's law: $i_b = \dfrac{2}{100} = 0.02 \text{ A}$

From Ohm's law: $v_2 = (-200i_b)(5) \Rightarrow \boxed{v_2 = -20 \text{ V}}$

15.

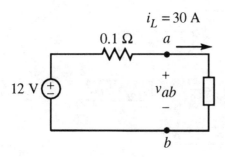

Assume that $i_L = 30$ A does not change.

KVL : $-12 + 0.1(30) + v_{ab} = 0 \Rightarrow \boxed{v_{ab} = 9 \text{ V}}$

In practical circuit, the load current would usually change if v_{ab} changed. Assume that the load is a resistance. From Example 2.11, for a load current of 30 A, $v_{ab} = 12.5$ V.

$$R_L = \frac{v_{ab}}{i_L} = \frac{12.5}{30} = 0.4167 \ \Omega$$

Then, from KVL: $-12 + 0.1\,i_L + 0.4167 i_L = 0 \Rightarrow \boxed{i_L = 23.22 \text{ A}}$

Finally, from Ohm's law: $\boxed{v_{ab} = R_L i_L = 9.68 \text{ V}}$

16. (a) $C = 0.01$ F, $v(t) = 100\cos(2\pi 60 t)$ V

$$i = C\frac{dv}{dt} = 0.01\frac{d}{dt}100\cos(2\pi 60 t) \Rightarrow \boxed{i = -377\sin(2\pi 60 t) \text{ A}}$$

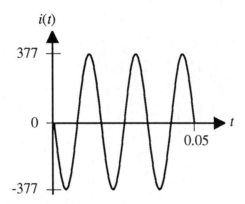

(b) $C = 0.01$ F, $v(t) = 100\cos(2\pi 60 t)$ V

From part (a), $i(t) = -377\sin(2\pi 60 t$ A$)$

$$p = vi = -37,700\cos(2\pi 60 t)\sin(2\pi 60 t) \Rightarrow \boxed{p = -18,850\sin(2\pi 120 t) \text{ W}}$$

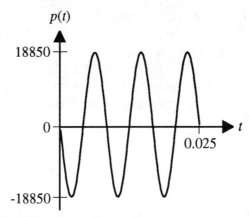

(c) $C = 0.01\ \text{F},\ v(t) = 100\cos(2\pi 60 t)\ \text{V}$

$$w(t) = \frac{1}{2}Cv^2(t) = \frac{1}{2}(0.01\ \text{F})\left[100\cos(2\pi 60 t)\ \text{V}\right]^2 = 50\cos^2(2\pi 60 t)\ \text{J}$$

$$\boxed{w(t) = 25 + 25\cos(2\pi 120 t)\ \text{J}}$$

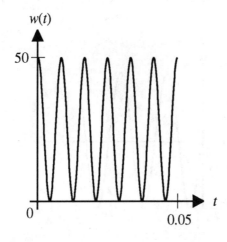

17.   Note that $v_a = 24e^{-3t}\ \text{V}$ and $v_b = 2e^{-3t}\ \text{V}$

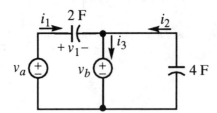

(a)   From KVL: $v_1 = v_a - v_b = 24e^{-3t} - 2e^{-3t}\ \Rightarrow\ \boxed{v_1 = 22e^{-3t}\ \text{V}}$

(b) $i_1 = 2\dfrac{dv_1}{dt} = 2\dfrac{d}{dt}\left(22e^{-3t}\right)\ \Rightarrow\ \boxed{i_1 = -132e^{-3t}\ \text{A}}$

(c) $i_2 = -4\dfrac{dv_b}{dt} = -4\dfrac{d}{dt}\left(2e^{-3t}\right)\ \Rightarrow\ \boxed{i_2 = 24e^{-3t}\ \text{A}}$

(d)   From KCL: $-i_1 + i_3 - i_2 = 0 \Rightarrow i_3 = -132e^{-3t} + 24e^{-3t}$

Therefore, $\boxed{i_3 = -108e^{-3t} \text{ A}}$

18.   Given $i_{ba} = -C\dfrac{d}{dt}\left(v_{ab}\right)$

$$\int_{-\infty}^{t} d\left[v_{ab}(\lambda)\right] = -\frac{1}{C}\int_{-\infty}^{t} i_{ba}(\lambda)d\lambda$$

$$v_{ab}(t) - v_{ab}(-\infty) = -\frac{1}{C}\int_{-\infty}^{t_0} i_{ba}(\lambda)d\lambda - \frac{1}{C}\int_{t_0}^{t} i_{ba}(\lambda)d\lambda$$

$$v_{ab}(t) = v_{ab}(-\infty) - \frac{1}{C}\int_{-\infty}^{t_0} i_{ba}(\lambda)d\lambda - \frac{1}{C}\int_{t_0}^{t} i_{ba}(\lambda)d\lambda$$

With the assumption that the capacitor voltage at $t = -\infty$ is zero, the equation above becomes

$$v_{ab}(t) = -\frac{1}{C}\int_{-\infty}^{t_0} i_{ba}(\lambda)d\tau - \frac{1}{C}\int_{t_0}^{t} i_{ba}(\lambda)d\lambda$$

Note that $-\dfrac{1}{C}\displaystyle\int_{-\infty}^{t_0} i_{ba}(\lambda)d\lambda = v_{ab}(t_0)$ by definition. Then,

$$v_{ab}(t) = v_{ab}(t_0) - \frac{1}{C}\int_{t_0}^{t} i_{ba}(\lambda)d\lambda$$

19.   (a) $L = 10$ mH and $i(t) = 100\cos(2\pi 60 t)$ A

$$v(t) = L\frac{di(t)}{dt} = 0.01\frac{d}{dt}\left[100\cos(2\pi 60 t)\right] = -120\pi\sin(2\pi 60 t)$$

$$\boxed{v(t) = -377\sin(2\pi 60 t) \text{ V}}$$

(b) $p(t) = v(t)i(t) = \left[-377\sin(2\pi 60 t)\text{ V}\right]\left[100\cos(2\pi 60 t)\text{ A}\right]$

$p(t) = -37700\sin(2\pi 60 t)\cos(2\pi 60 t) \Rightarrow \boxed{p(t) = -18,850\sin(2\pi 120 t) \text{ W}}$

(c) $w(t) = \dfrac{1}{2}Li^2(t) = \dfrac{1}{2}(0.01)\left[100\cos(2\pi 60 t)\right]^2 = 50\cos^2(2\pi 60 t)$ J

$$\boxed{w(t) = 25 + 25\cos(2\pi 120 t) \text{ J}}$$

20.   Note that $i_a = 24e^{-3t}$ A and $i_b = 2e^{-3t}$ A.

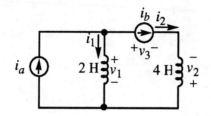

(a)   From KCL: $-i_a + i_1 + i_b = 0 \implies i_1 = 24e^{-3t} - 2e^{-3t}$

$$\boxed{i_1 = 22e^{-3t} \ A}$$

(b) $i_2 = i_b \implies \boxed{i_2 = 2e^{-3t} \ A}$

(c) $v_1 = 2\dfrac{di_1}{dt} = 2\dfrac{d}{dt}\left(22e^{-3t}\right) \implies \boxed{v_1 = -132e^{-3t} \ V}$

(d) $v_2 = -4\dfrac{di_2}{dt} = -4\dfrac{d}{dt}\left(2e^{-3t}\right) \implies \boxed{v_2 = 24e^{-3t} \ V}$

21.   Given $v_{ba}(t) = -L\dfrac{d}{dt}\left(i_{ab}\right)$

$$\int_{-\infty}^{t} d\left[i_{ab}(\lambda)\right] = -\frac{1}{L}\int_{-\infty}^{t} v_{ba}(\lambda)d\lambda$$

$$i_{ab}(t) - i_{ab}(-\infty) = -\frac{1}{L}\int_{-\infty}^{t_0} v_{ba}(\lambda)d\lambda - \frac{1}{L}\int_{t_0}^{t} v_{ba}(\lambda)d\lambda$$

$$i_{ab}(t) = i_{ab}(-\infty) - \frac{1}{L}\int_{-\infty}^{t} v_{ba}(\lambda)d\lambda - \frac{1}{L}\int_{t_0}^{t} v_{ba}(\lambda)d\lambda$$

If the inductor current at $t = -\infty$ is zero, the equation above becomes

$$i_{ab}(t) = -\frac{1}{L}\int_{-\infty}^{t_0} v_{ba}(\lambda)d\lambda - \frac{1}{L}\int_{t_0}^{t} v_{ba}(\lambda)d\lambda$$

Note that $-\dfrac{1}{L}\displaystyle\int_{-\infty}^{t_0} v_{ba}(\lambda)d\lambda = i_{ab}(t_0)$ by definition. Then,

$$i_{ab}(t) = i_{ab}(t_0) - \frac{1}{L}\int_{0}^{t} v_{ba}(\lambda)d\lambda$$

CHAPTER 3

Introduction to Circuit Analysis

Exercises

1.   Given the circuit shown below:

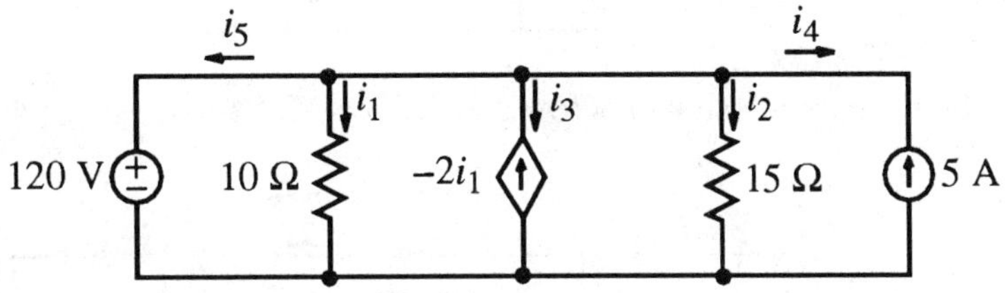

$$i_1 = \frac{v}{R} = \frac{120 \text{ V}}{10 \text{ }\Omega} \Rightarrow \boxed{i_1 = 12 \text{ A}} \quad : \quad i_2 = \frac{v}{R} = \frac{120 \text{ V}}{15 \text{ }\Omega} \Rightarrow \boxed{i_2 = 8 \text{ A}}$$

$$i_3 = -\left(-2i_1\right) \Rightarrow \boxed{i_3 = 24 \text{ A}} \quad : \quad \boxed{i_4 = -5 \text{ A}}$$

From KCL: $i_5 = -i_1 - i_2 - i_3 - i_4 = -12 - 8 - 24 + 5 \Rightarrow \boxed{i_5 = -39 \text{ A}}$

$$p_1 = vi_1 = (120 \text{ V})(12 \text{ A}) \Rightarrow \boxed{p_1 = 1440 \text{ W}}$$

$$p_2 = vi_2 = (120 \text{ V})(8 \text{ A}) \Rightarrow \boxed{p_2 = 960 \text{ W}}$$

$$p_3 = vi_3 = (120 \text{ V})(24 \text{ A}) \Rightarrow \boxed{p_3 = 2880 \text{ W}}$$

$$p_4 = vi_4 = (120 \text{ V})(-5 \text{ A}) \Rightarrow \boxed{p_4 = -600 \text{ W}}$$

$$p_5 = vi_5 = (120 \text{ V})(-39 \text{ A}) \Rightarrow \boxed{p_5 = -4680 \text{ W}}$$

Note that $p_1 + p_2 + p_3 + p_4 + p_5 = 0$

2.   Given the circuit shown below (the voltage v is not known)

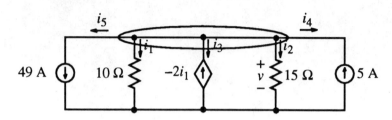

$$i_1 = \frac{v}{10} \ : \ i_2 = \frac{v}{15} \ : \ i_3 = -(-2i_1) = \frac{v}{5} \ : \ i_4 = -5\,A \ : \ i_5 = 49\,A$$

From KCL: $\ i_1 + i_2 + i_3 + i_4 + i_5 = 0$

Then, $\ \dfrac{v}{10} + \dfrac{v}{15} + \dfrac{v}{5} - 5\,A + 49\,A = 0 \ \Rightarrow \ \boxed{v = -120\,V}$

Therefore,

$$i_1 = -\frac{-120\,V}{10\,\Omega} \ \Rightarrow \ \boxed{i_1 = -12\,A} \ : \ i_2 = \frac{-120\,V}{15\,\Omega} \ \Rightarrow \ \boxed{i_2 = -8\,A}$$

$$i_3 = -(-2i_1) \ \Rightarrow \ \boxed{i_3 = -24\,A} \ , \ \boxed{i_4 = -5\,A} \ , \ \boxed{i_5 = 49\,A}$$

3.   Given the circuit shown below and $i_3(0) = -5\,A$:

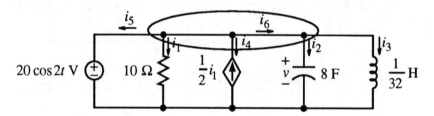

$$i_1 = \frac{v}{10\,\Omega} = \frac{20\cos(2t)\,V}{10\,\Omega} \ \Rightarrow \ \boxed{i_1 = 2\cos(2t)\,A}$$

$$i_2 = C\frac{dv}{dt} = 8\frac{d}{dt}[20\cos(2t)] \ \Rightarrow \ \boxed{i_2 = -320\sin(2t)\,A}$$

$$i_3 = i_3(0) + \frac{1}{L}\int_0^t v(\lambda)d\lambda = -5 + 32\int_0^t 20\cos(2\lambda)d\lambda$$

$$\boxed{i_3 = -5 + 320\sin(2t)\,A} \ , \ i_4 = -\left(\frac{1}{2}i_1\right) \ \Rightarrow \ \boxed{i_4 = -\cos(2t)\,A}$$

From KCL: $\ i_5 = -i_1 - i_2 - i_3 - i_4$

$$= -2\cos(2t) + 320\sin(2t) + 5 - 320\sin(2t) + \cos(2t)$$

$$\boxed{i_5 = 5 - \cos(2t)\,A}$$

From KCL: $\ i_6 = i_2 + i_3 = -320\sin(2t) - 5 + 320\sin(2t) \ \Rightarrow \ \boxed{i_6 = -5\,A}$

4.   Given the circuit shown below with $v(0) = 20$ V and $i_3 = (0) = 10$ A:

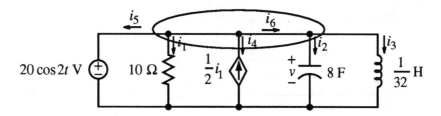

From KCL: $i_1 + i_2 + i_3 + i_4 + i_5 = 0$

Using the component equations gives the following

$$\frac{v}{10} + 8\frac{dv}{dt} + 10 + 32\int_0^t v(\lambda)d\lambda - \frac{1}{2}\left(\frac{v}{10}\right) - 2\cos(2t) = 0$$

$$\boxed{8\frac{dv}{dt} + \frac{v}{20} + 32\int_0^t v(\lambda)d\lambda = -10 + 2\cos(2t)}$$

5.   (a)   Given the circuit below, and $i_{ab} = e^{2t}$

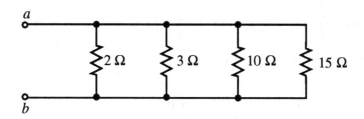

$$R_{ab} = \frac{1}{\dfrac{1}{2} + \dfrac{1}{3} + \dfrac{1}{10} + \dfrac{1}{15}} \quad \Rightarrow \quad \boxed{R_{ab} = 1\,\Omega} \; : \; \boxed{v_{ab} = Ri_{ab} = e^{2t}\ \text{V}}$$

(b)   Given the circuit below, with the resistors all in parallel, and $i_{ab} = e^{2t}$ A:

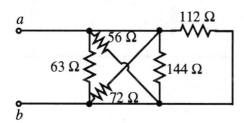

$$R_{ab} = \frac{1}{\dfrac{1}{112} + \dfrac{1}{144} + \dfrac{1}{72} + \dfrac{1}{56} + \dfrac{1}{63}} \quad \Rightarrow \quad \boxed{R_{ab} = 15.75\,\Omega}$$

$$v_{ab} = R_{ab}i_{ab} = (15.75\,\Omega)(e^{2t}) \quad \Rightarrow \quad \boxed{v = 15.75e^{2t}\ \text{V}}$$

(c)   Given the circuit below (the inductors are all in parallel):

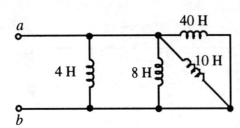

$$L_{ab} = \frac{1}{\frac{1}{40} + \frac{1}{10} + \frac{1}{8} + \frac{1}{4}} \Rightarrow \boxed{L_{ab} = 2\,\text{H}}$$

$$v_{ab} = L_{ab}\frac{di_{ab}}{dt} = 2\frac{d}{dt}\left(e^{2t}\right) \Rightarrow \boxed{v_{ab} = 4e^{2t}\,\text{V}}$$

(d)   Given the circuit shown below (the capacitors are all in parallel):

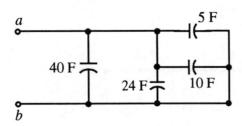

$$C_{ab} = 5 + 10 + 24 + 40 \Rightarrow \boxed{C_{ab} = 79\,\text{F}}$$

$$v_{ab} = \frac{1}{C_{ab}}\int_{-\infty}^{t} i_{ab}(\lambda)d\lambda = \frac{1}{79}\int_{-\infty}^{t} e^{2\lambda}d\lambda = \frac{1}{158}e^{2\lambda}\Big|_{-\infty}^{t} \Rightarrow \boxed{v_{ab} = 6.329e^{2t}\,\text{mV}}$$

6.

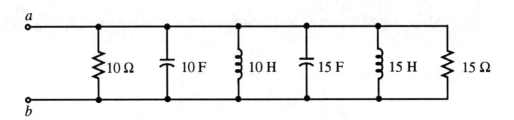

$$R_p = \frac{1}{\frac{1}{10} + \frac{1}{15}} = 6\,\Omega$$

$$L_p = \frac{1}{\frac{1}{10} + \frac{1}{15}} = 6\,\text{H}$$

$$C_p = 10 + 15 = 25\,\text{F}$$

$$i = \frac{1}{R}v + \frac{1}{L}\int_{-\infty}^{t} v \, d\lambda + C\frac{d}{dt}v$$

$$= \frac{1}{6}360e^{2t} + \frac{1}{6}\int_{-\infty}^{t} 360e^{2\lambda} \, d\lambda + 25\frac{d}{dt}360e^{2t}$$

$$= (60 + 30 + 18,000)e^{2t} \Rightarrow \boxed{i = 18,090e^{2t} \text{ A}}$$

7.   (a)   Given the circuit shown below:

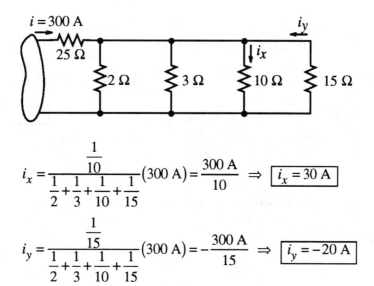

$$i_x = \frac{\dfrac{1}{10}}{\dfrac{1}{2} + \dfrac{1}{3} + \dfrac{1}{10} + \dfrac{1}{15}}(300 \text{ A}) = \frac{300 \text{ A}}{10} \Rightarrow \boxed{i_x = 30 \text{ A}}$$

$$i_y = \frac{\dfrac{1}{15}}{\dfrac{1}{2} + \dfrac{1}{3} + \dfrac{1}{10} + \dfrac{1}{15}}(300 \text{ A}) = -\frac{300 \text{ A}}{15} \Rightarrow \boxed{i_y = -20 \text{ A}}$$

    (b)   Given the circuit shown below

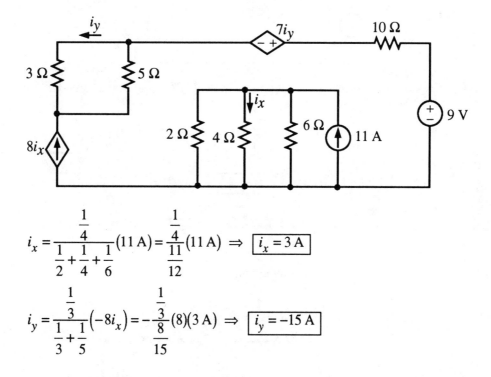

$$i_x = \frac{\dfrac{1}{4}}{\dfrac{1}{2} + \dfrac{1}{4} + \dfrac{1}{6}}(11 \text{ A}) = \frac{\dfrac{1}{4}}{\dfrac{11}{12}}(11 \text{ A}) \Rightarrow \boxed{i_x = 3 \text{ A}}$$

$$i_y = \frac{\dfrac{1}{3}}{\dfrac{1}{3} + \dfrac{1}{5}}(-8i_x) = -\frac{\dfrac{3}{8}}{\dfrac{15}{8}}(8)(3 \text{ A}) \Rightarrow \boxed{i_y = -15 \text{ A}}$$

8.   Given the circuit shown below:

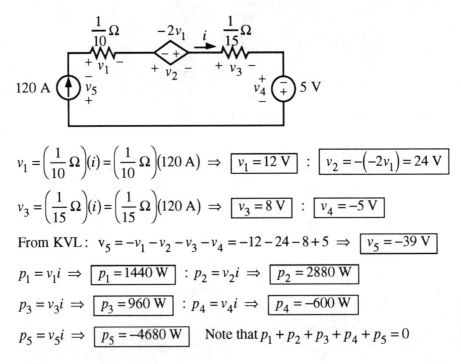

$$v_1 = \left(\frac{1}{10}\,\Omega\right)(i) = \left(\frac{1}{10}\,\Omega\right)(120\text{ A}) \;\Rightarrow\; \boxed{v_1 = 12\text{ V}} \;:\; \boxed{v_2 = -(-2v_1) = 24\text{ V}}$$

$$v_3 = \left(\frac{1}{15}\,\Omega\right)(i) = \left(\frac{1}{15}\,\Omega\right)(120\text{ A}) \;\Rightarrow\; \boxed{v_3 = 8\text{ V}} \;:\; \boxed{v_4 = -5\text{ V}}$$

From KVL: $v_5 = -v_1 - v_2 - v_3 - v_4 = -12 - 24 - 8 + 5 \;\Rightarrow\; \boxed{v_5 = -39\text{ V}}$

$p_1 = v_1 i \;\Rightarrow\; \boxed{p_1 = 1440\text{ W}} \;:\; p_2 = v_2 i \;\Rightarrow\; \boxed{p_2 = 2880\text{ W}}$

$p_3 = v_3 i \;\Rightarrow\; \boxed{p_3 = 960\text{ W}} \;:\; p_4 = v_4 i \;\Rightarrow\; \boxed{p_4 = -600\text{ W}}$

$p_5 = v_5 i \;\Rightarrow\; \boxed{p_5 = -4680\text{ W}}$ Note that $p_1 + p_2 + p_3 + p_4 + p_5 = 0$

9.   Given the circuit below:

The current i is unknown, so write a KVL equation around the loop.

$$49 + \left(\frac{1}{10}\,\Omega\right)(i) - \left[-2\left(\frac{1}{10}\right)(i)\right] + \frac{1}{15}(i) - 5 = 0 \;:\; \boxed{i = \left(\frac{30}{11}\right)(-44) = -120\text{ A}}$$

$$v_1 = \left(\frac{1}{10}\,\Omega\right)(i) \;\Rightarrow\; \boxed{v_1 = -12\text{ V}} \;:\; v_2 = -(-2v_1) \;\Rightarrow\; \boxed{v_2 = -24\text{ V}}$$

$$v_3 = \left(\frac{1}{15}\,\Omega\right)(i) \;\Rightarrow\; \boxed{v_3 = -8\text{ V}} \;:\; \boxed{v_4 = -5\text{ V}} \;:\; \boxed{v_5 = 49\text{ V}}$$

10.   Given the circuit shown below, and $v_3(0) = -5$ V :

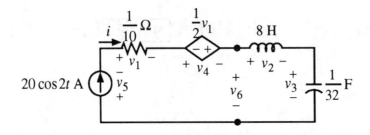

$$v_1 = \left(\frac{1}{10}\,\Omega\right)(i) = \left(\frac{1}{10}\,\Omega\right)[20\cos(2t)] \;\Rightarrow\; \boxed{v_1 = 2\cos(2t)\text{ V}}$$

$$v_2 = 8\frac{d}{dt}(i) = 8\frac{d}{dt}[20\cos(2t)] \;\Rightarrow\; \boxed{v_2 = -320\sin(2t)\text{ V}}$$

$$v_3 = v_3(0) + 32\int_0^t i(\lambda)d\lambda = -5 + 32\int_0^t 20\cos(2\lambda)d\lambda = -5 + 320\sin(2\lambda)\big|_0^t$$

$$\boxed{v_3 = -5 + 320\sin(2t)\text{ V}} \;:\; v_4 = -\frac{1}{2}(v_1) \;\Rightarrow\; \boxed{v_4 = -\cos(2t)\text{ V}}$$

From KVL: $v_5 = -v_1 - v_2 - v_3 - v_4 \;\Rightarrow\; \boxed{v_5 = 5 - \cos(2t)\text{ V}}$

From KVL: $v_6 = v_2 + v_3 \;\Rightarrow\; \boxed{v_6 = -5\text{ V}}$

11.   Given the circuit shown below, $i(0) = 20$ A, and $v_3(0) = 10$ V:

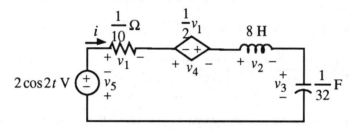

From KVL: $v_1 + v_2 + v_3 + v_4 + v_5 = 0$

Using the following component equations gives the following

$$\frac{i}{10} + 8\frac{di}{dt} + 10 + 32\int_0^t i(\lambda)d\lambda - \frac{1}{2}\left(\frac{i}{10}\right) - 2\cos(2t) = 0$$

$$\boxed{8\frac{di}{dt} + \frac{i}{20} + 32\int_0^t i(\lambda)d\lambda = -10 + 2\cos(2t)}$$

12.   (a)   Given the circuit shown below and $v_{ab} = 148e^{3t}$ V:

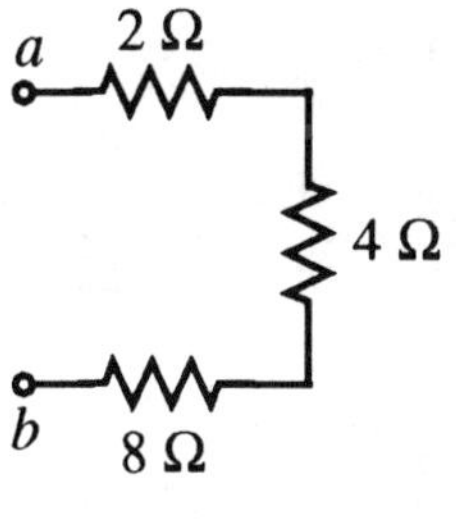

a 2 Ω

4 Ω

b 8 Ω

$$\boxed{R_{ab} = 2 + 4 + 8 = 14\ \Omega} \;:\; i_{ab} = \frac{v_{ab}}{R_{ab}} = \frac{148e^{3t}\text{ V}}{14\ \Omega} \;\Rightarrow\; \boxed{i_{ab} = 10.57e^{3t}\text{ A}}$$

NOTE. For ease of notation, the symbol ‖ will denote "in parallel with." For example

$$R_1 \| R_2 = \cfrac{1}{\cfrac{1}{R_1}+\cfrac{1}{R_2}} = \frac{R_1 R_2}{R_1 + R_2}$$

(b)   Given the circuit shown below and $v_{ab} = 148e^{3t}$ V:

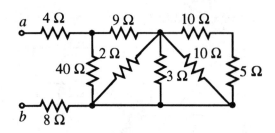

$$R_{s1} = 10+5 = 15\,\Omega \ : \ R_{p1} = R_{s1} \| 10 \| 3 \| 2 = 15 \| 10 \| 3 \| 2 = 1\,\Omega$$

$$R_{s2} = R_{p1} + 9 = 1+9 = 10\,\Omega \ : \ R_{p2} = R_{s2} \| 40 = 10 \| 40 = 8\,\Omega$$

$$R_{ab} = 4 + R_{s2} + 8 = 4+8+8 \ \Rightarrow \ \boxed{R_{ab} = 20\,\Omega}$$

$$i_{ab} = \frac{v_{ab}}{R_{ab}} = \frac{148e^{3t}\ \text{V}}{20\,\Omega} \ \Rightarrow \ \boxed{i_{ab} = 7.4e^{3t}\ \text{A}}$$

(c)   Given the circuit shown below and $v_{ab} = 148e^{3t}$ V:

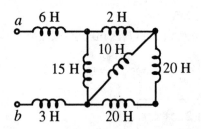

$$L_{s1} = 20+20 = 40\,\text{H} \ : \ L_{p1} = L_{s1} \| 10 = 40 \| 10 = 8\,\text{H}$$

$$L_{s2} = L_{p1} + 2 = 8+2 = 10\,\text{H} \ : \ L_{p2} = L_{s2} \| 15\,\text{H} = 10 \| 15 = 6\,\text{H}$$

$$L_{ab} = 6 + L_{p2} + 3 = 6+6+3 \ \Rightarrow \ \boxed{L_{ab} = 15\,\text{H}}$$

$$i_{ab} = \frac{1}{L_{ab}} \int_{-\infty}^{t} v_{ab}(\lambda)d\lambda = \frac{1}{15}\int_{-\infty}^{t} 148e^{3\lambda}d\lambda = \frac{148}{45}e^{3\lambda}\Big|_{-\infty}^{t} \ \Rightarrow \ \boxed{i_{ab} = 3.29e^{3t}\ \text{A}}$$

(d)   Given the circuit shown below and $v_{ab} = 148e^{3t}$ V:

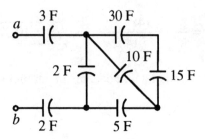

$$C_{s1} = \frac{1}{\frac{1}{15} + \frac{1}{30}} = 10\,\text{F}$$

$$C_{p1} = 10 + 10 = 20\,\text{F}$$

$$C_{s2} = \frac{1}{\frac{1}{5} + \frac{1}{20}} = 4\,\text{F}$$

$$C_{p2} = 2 + 4 = 6\,\text{F}$$

$$C = \frac{1}{\frac{1}{3} + \frac{1}{6} + \frac{1}{2}} = \quad \Rightarrow \quad \boxed{C = 1\,\text{F}}$$

$$i_{ab} = C\frac{d}{dt}v_{ab} = 1\frac{d}{dt}\left(148e^{3t}\right) \quad \Rightarrow \quad \boxed{i = 444e^{3t}\ \text{V}}$$

13.

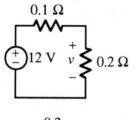

$$v = \frac{0.2}{0.1 + 0.2}(12\,\text{V}) \quad \Rightarrow \quad \boxed{v = 8\,\text{V}}$$

14.   (a)   Given the circuit shown below:

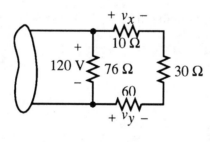

$$v_x = \frac{10}{10 + 30 + 60}(120\,\text{V}) \quad \Rightarrow \quad \boxed{v_x = 12\,\text{V}}$$

$$v_y = -\frac{60}{10 + 30 + 60}(120\,\text{V}) \quad \Rightarrow \quad \boxed{v_y = -72\,\text{V}}$$

(b)   Given the circuit below:

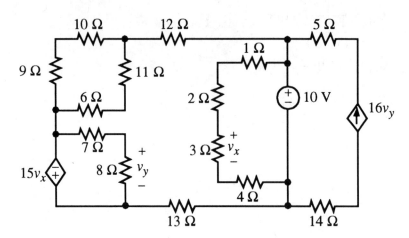

$$v_x = \frac{3}{1+2+3+4}(10 \text{ V}) = 3 \text{ V}$$

$$v_y = -\frac{8}{7+8}(15v_x) = -\frac{8}{15}(15)(3) \Rightarrow \boxed{v_y = -24 \text{ V}}$$

15.   (a)   Given the circuit shown below:

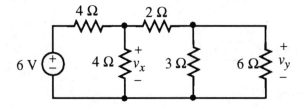

This circuit can be reduced by combining the 2-Ω, 3-Ω, and 6-Ω resistors as shown

$$R_1 = 2 + (3 \| 6) = 4 \ \Omega$$

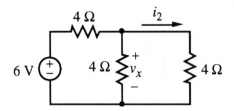

Then this resistance can be combined in parallel with the 4 Ω. The circuit then becomes

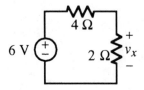

$$v_x = \frac{2}{4+2}(6\text{ V}) \Rightarrow \boxed{v_x = 2\text{ V}}$$

From the previous circuit, $i_2 = \frac{1}{4}\left(v_x\right) = \frac{1}{2}\text{ A}$

$$i_y = \frac{3}{3+6}\left(i_2\right) = \frac{1}{6}\text{ A} \;:\; v_y = 6i_y \Rightarrow \boxed{v_y = 1\text{ V}}$$

(b) Given the circuit shown below:

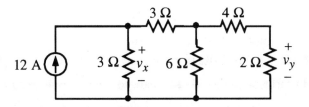

Combining the 4 Ω and the 2 Ω in series yields 6 Ω.

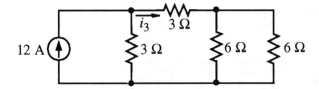

Then the two 6-Ω resistors can be combined in parallel to give 3-Ω.

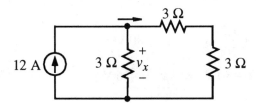

The two 3-Ω resistors canbe combined in series to yield 6 Ω.

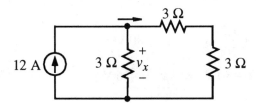

$$\boxed{v_x = 12\text{ A}(3\,\Omega \,\|\, 6\,\Omega) = 24\text{ V}} \;:\; i_3 = \frac{v_x}{6\,\Omega} = 4\text{ A}$$

Use i_3 and current divider in the first equivalent circuit to give:

$$i_y = \frac{6}{6+6}\left(i_3\right) = 2\text{ A} \;:\; v_y = 2i_y \Rightarrow \boxed{v_y = 4\text{ V}}$$

(c)   Given the circuit shown below:

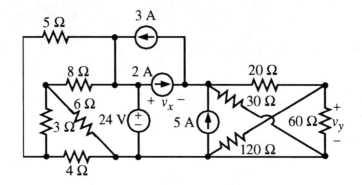

The circuit can be simplified by the following steps.

Combine the 60 Ω and the 120Ω in parallel to give $R_p = 60\,\Omega \,||\, 120\,\Omega = 40\,\Omega$.

Combine R_p in series with the 20 Ω resistor to give $R_s = 40 + 20 = 60\,\Omega$.

Finally, R_s is in parallel with the 30-Ω resistor so that $R_{p2} = 60\,||\,30 = 20\,\Omega$

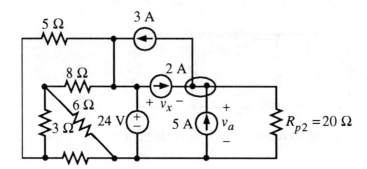

From KCL: $3 - 2 - 5 + \dfrac{v_a}{20} = 0 \;\Rightarrow\; v_a = 80\text{ V}$

From KVL: $v_x = 24\text{ V} - 80\text{ V} \;\Rightarrow\; \boxed{v_x = -56\text{ V}}$

Referring to the original circuit and using a voltage divider yields:

$$v_y = \frac{R_p}{R_p + 20}\left(v_a\right) = \frac{40}{40 + 20}(80\text{ V}) \;\Rightarrow\; \boxed{v_y = 53.33\text{ V}}$$

CHAPTER 4

Node Voltages and Mesh Currents

Exercises

1.   Given the circuit shown below:

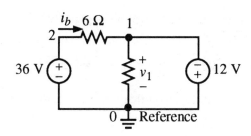

Node 1, KVL : $\boxed{v_1 = -12 \text{ V}}$ Node 2, KVL : $\boxed{v_2 = 36 \text{ V}}$

Then v_{12} is equal to the difference of the node voltages

$$\boxed{v_{12} = v_1 - v_2 = -48 \text{ V}}$$

$$i_b = -i_{12} = -\left(\frac{1}{6}\right)v_{12} \Rightarrow \boxed{i_{12} = 8 \text{ A}}$$

2.   Given the circuit shown below:

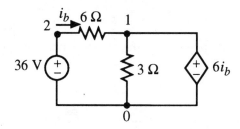

Node 1, KVL : $v_1 = 6i_b = 6\left[\frac{1}{6}(v_2 - v_1)\right] = v_2 - v_1 \Rightarrow \boxed{v_2 = 36 \text{ V}} \Rightarrow \boxed{v_1 = 18 \text{ V}}$

$\boxed{v_{12} = v_1 - v_2 = -18 \text{ V}}$ and $i_b = -i_{12} = -\left(\frac{1}{6}\right)v_{12} \Rightarrow \boxed{i_b = 3 \text{ A}}$

3.   (a)   Given the circuit shown below:

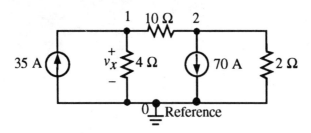

Node 1, KCL: $-35 + \frac{1}{4}v_1 + \frac{1}{10}(v_1 - v_2) = 0$

Node 2, KCL: $\frac{1}{10}(v_2 - v_1) + 70 + \frac{1}{2}v_2 = 0$

These two equations yield the following matrix equation:

$$\begin{bmatrix} \frac{7}{20} & -\frac{1}{10} \\ -\frac{1}{10} & \frac{3}{5} \end{bmatrix} \cdot \begin{bmatrix} v_1 \\ v_2 \end{bmatrix} = \begin{bmatrix} 35 \\ -70 \end{bmatrix}$$

Cramer's rule, a calculator, or a computer can be used to solve this equtaion.
$\boxed{v_1 = 70 \text{ V}}$ and $\boxed{v_2 = -105 \text{ V}}$. The voltage v_{12} can be found from the
difference of the two node voltages. $\boxed{v_{12} = v_1 - v_2 = 175 \text{ V}}$

(b)   Refer to the circuit shown in part (a). Application of KCL to the reference node
yields

$$35 - \frac{1}{4}v_1 - 70 - \frac{1}{2}v_2 = 0 \implies -\frac{1}{4}v_1 - \frac{1}{2}v_2 = 35$$

Add the two KCL equations found in part (a):

Node 1, KCL: $-35 + \frac{1}{4}v_1 + \frac{1}{10}(v_1 - v_2) = 0$

Node 2, KCL: $\frac{1}{10}(v_2 - v_1) + 70 + \frac{1}{2}v_2 = 0$

$$\frac{1}{4}v_1 + \frac{1}{2}v_2 = -35$$

Notice that this equation is the negative of the equation obtained by a KCL at the
reference node.

4.   Given the circuit shown below:

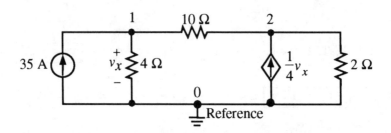

Observe that $v_x = v_1$. Then sum the currents leaving nodes 1 and 2.

Node 1, KCL: $-35 + \dfrac{1}{4}v_1 + \dfrac{1}{10}\left(v_1 - v_2\right) = 0$

Node 2, KCL: $\dfrac{1}{10}\left(v_2 - v_1\right) - \dfrac{1}{4}v_1 + \dfrac{1}{2}v_2 = 0$

$$\begin{bmatrix} \frac{7}{20} & -\frac{1}{10} \\ -\frac{7}{20} & \frac{3}{5} \end{bmatrix} \cdot \begin{bmatrix} v_1 \\ v_2 \end{bmatrix} = \begin{bmatrix} 35 \\ 0 \end{bmatrix} \quad \boxed{v_1 = 120 \text{ V}} \text{ and } \boxed{v_2 = 70 \text{ V}}$$

$v_{12} = v_1 - v_2 \implies \boxed{v_{12} = 50 \text{ V}}$

5. Given the circuit below:

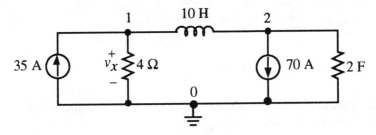

Node 1, KCL: $-35 + \dfrac{1}{4}v_1 + \dfrac{1}{10}\displaystyle\int_{-\infty}^{t}\left(v_1 - v_2\right)d\lambda = 0$

Node 2, KCL: $\dfrac{1}{10}\displaystyle\int_{-\infty}^{t}\left(v_2 - v_1\right)d\lambda + 70 + 2\dfrac{d}{dt}v_2 = 0$

These equations can be written in matrix form using the notation

$$\int_{-\infty}^{t} v\, d\lambda \triangleq \int_{-\infty}^{t} d\lambda\, v$$

These equations in matrix form are shown below:

$$\begin{bmatrix} \frac{1}{4} + \frac{1}{10}\int_{-\infty}^{t} d\lambda & -\frac{1}{10}\int_{-\infty}^{t} d\lambda \\ -\frac{1}{10}\int_{-\infty}^{t} d\lambda & 2\frac{d}{dt} + \frac{1}{10}\int_{-\infty}^{t} d\lambda \end{bmatrix} \cdot \begin{bmatrix} v_1 \\ v_2 \end{bmatrix} = \begin{bmatrix} 35 \\ -70 \end{bmatrix}$$

6. (a) Given the circuit below:

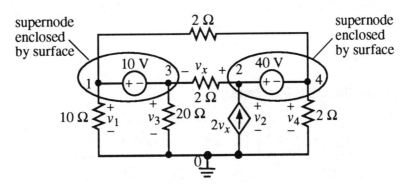

Node 1- 3, KCL: $\frac{1}{10}v_1 + \frac{1}{2}(v_1 - v_4) + \frac{1}{20}v_3 + \frac{1}{2}(v_3 - v_2) = 0$

$\Rightarrow \quad \frac{3}{5}v_1 - \frac{1}{2}v_2 + \frac{11}{20}v_3 - \frac{1}{2}v_4 = 0$

Node 1- 3, KVL: $v_1 - v_3 = 10$

Node 2 - 4, KCL: $\frac{1}{2}(v_2 - v_3) - 2(v_2 - v_3) + \frac{1}{2}(v_4 - v_1) + \frac{1}{2}v_4 = 0$

$\Rightarrow \quad -\frac{1}{2}v_1 - \frac{3}{2}v_2 + \frac{3}{2}v_3 + v_4 = 0$

Node 2 - 4, KVL: $v_2 - v_4 = 40$

These equations can be solved for the four node voltages. The algebra can be simplified by writing the equations in terms of only v_1 and v_2, using the two KVL equations to solve for v_3 and v_4 in terms of v_1 and v_2.

From KVL: $v_3 = v_1 - 10$ and $v_4 = v_2 - 40$

Substituting these two equations into the two KCL equations yields two equations and two unknowns.

Node 1- 3, KCL: $\frac{1}{10}v_1 + \frac{1}{2}(v_1 - v_2 + 40) + \frac{1}{20}(v_1 - 10) + \frac{1}{2}(v_1 - v_2 - 10) = 0$

$\Rightarrow \quad \boxed{\frac{23}{20}v_1 - v_2 = -\frac{29}{2}}$

Node 2 - 4, KCL: $\frac{1}{2}(v_2 - v_1 + 10) - 2(v_2 - v_1 + 10) + \frac{1}{2}(v_2 - v_1 - 40)$

$+ \frac{1}{2}(v_2 - 40) = 0$

$\boxed{v_1 - \frac{1}{2}v_2 = 55}$

Solving these four equations yields:

$v_1 = 146.47$ V

$v_2 = 182.94$ V

$v_3 = 136.47$ V

$v_4 = 142.94$ V

(b)   Using the results from part (a) ,

$\boxed{i_{10} = \frac{1}{10}v_1 = 14.65 \text{ A}}$: $\boxed{i_{14} = \frac{1}{2}(v_1 - v_4) = 1.765 \text{ A}}$

Node 1, KCL: $i_{13} = -i_{10} - i_{14} \Rightarrow \boxed{i_{13} = -16.41 \text{ A}}$

7.   Given the circuit shown below:

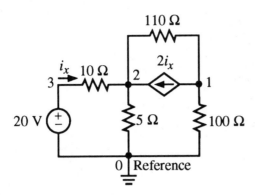

Node 1, KCL: $\dfrac{1}{110}(v_1 - v_2) + 2\left(\dfrac{1}{10}\right)(20 - v_2) + \dfrac{1}{100}v_1 = 0$

Node 2, KCL: $\dfrac{1}{10}(v_2 - 20) + \dfrac{1}{5}v_2 + \dfrac{1}{110}(v_2 - v_1) - 2\dfrac{1}{10}(20 - v_2) = 0$

$$\begin{bmatrix} \dfrac{21}{1100} & -\dfrac{23}{110} \\ -\dfrac{1}{110} & \dfrac{56}{110} \end{bmatrix} \cdot \begin{bmatrix} v_1 \\ v_2 \end{bmatrix} = \begin{bmatrix} -4 \\ 6 \end{bmatrix} \Rightarrow \boxed{\begin{aligned} v_1 &= -100 \text{ V} \\ v_2 &= 10 \text{ V} \end{aligned}} \quad \text{From KVL, } \boxed{v_3 = 20 \text{ V}}$$

The power absorbed by the dependent source is

$$p = v_{12}(2i_x) = (v_1 - v_2)(2)\left(\dfrac{1}{10}\right)(v_3 - v_2) \Rightarrow \boxed{p = -220 \text{ W}}$$

8.   Given the circuit shown below:

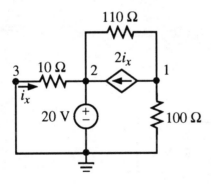

Node 1, KCL: $\dfrac{1}{110}(v_1 - 20) + 2\left[-\dfrac{1}{10}(20)\right] + \dfrac{1}{100}v_1 = 0 \Rightarrow \boxed{v_1 = 219 \text{ V}}$

Node 2, KVL: $\boxed{v_2 = 20 \text{ V}}$: Node 3, KVL: $\boxed{v_3 = 0}$

The power absorbed by the dependent source is

$$p = v_{12}(2i_x) = (v_1 - v_2)(2)\left(-\dfrac{20}{10}\right) \Rightarrow \boxed{p = -796 \text{ W}}$$

9.   Given the circuit shown below:

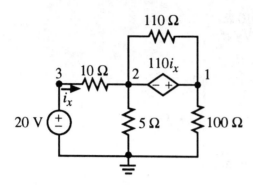

Node 1-2, KVL: $v_2 = v_1 - 110i_x = -110\left[\dfrac{1}{10}(20 - v_2)\right] + v_1$

$\Rightarrow \quad v_2 = 22 - \dfrac{1}{10}v_1$

Node 1-2, KCL: $\dfrac{1}{10}\left[\left(22 - \dfrac{1}{10}v_1\right) - 20\right] + \dfrac{1}{5}\left(22 - \dfrac{1}{10}v_1\right) + \dfrac{1}{100}v_1 = 0$

$\Rightarrow \quad \boxed{v_1 = 230\text{ V}}$

$v_2 = 22 - \dfrac{1}{10}v_1 \Rightarrow \boxed{v_2 = -1\text{ V}}$

The power in the dependent source is found by $p = v_{12}i_y$, which means that the current i_y must be found first.

Node 1, KCL: $i_y + \dfrac{1}{110}(v_1 - v_2) + \dfrac{1}{100}v_1 = 0 \Rightarrow \boxed{i_y = -4.4\text{ A}}$

$p = (v_1 - v_2)(i_y) \Rightarrow \boxed{p = -1016.4\text{ W}}$

10.   (a)   Given the circuit shown below

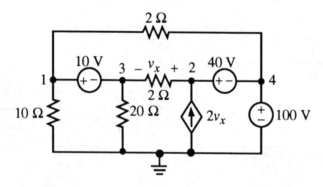

Node 1-3, KCL: $\dfrac{1}{10}v_1 + \dfrac{1}{2}(v_1 - v_4) + \dfrac{1}{20}v_3 + \dfrac{1}{2}(v_3 - v_2) = 0$

Node 1-3, KVL: $v_1 - v_3 = 10$: Node 2-4, KVL: $v_2 = 40 + 100 = 140$

Node 4, KVL: $v_4 = 100$

These equations are given in matrix form below:

$$\begin{bmatrix} \frac{3}{5} & -\frac{1}{2} & \frac{11}{20} & -\frac{1}{2} \\ 1 & 0 & -1 & 0 \\ 0 & 1 & 0 & 0 \\ 0 & 0 & 0 & 1 \end{bmatrix} \begin{bmatrix} v_1 \\ v_2 \\ v_3 \\ v_4 \end{bmatrix} = \begin{bmatrix} 0 \\ 10 \\ 140 \\ 100 \end{bmatrix}$$

(b)   Using the shortcut procedure, part (a) can be solved by using teh KVL equations substituted directly into the KCL equation.

From KVL : $\boxed{v_2 = 140 \text{ V}}$ $\boxed{v_4 = 100 \text{ V}}$ and $v_3 = v_1 - 10$

Node 1-3, KCL : $\dfrac{1}{10}v_1 + \dfrac{1}{2}(v_1 - 100) + \dfrac{1}{20}(v_1 - 10) + \dfrac{1}{2}[(v_1 - 10) - 140] = 0$

Notice that this equation invovles only one unknown, v_1, and can therefore be solved rather simply.

$\boxed{v_1 = 109.13 \text{ V}}$ Then from KVL : $\boxed{v_3 = 99.13 \text{ V}}$

11.   Given the circuit shown below:

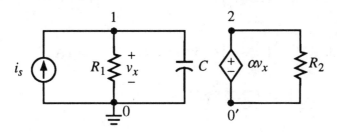

Node 1, KCL : $-i_s + \dfrac{v_1}{R_1} + C\dfrac{d}{dt}v_1 = 0$ Node 2, KVL : $v_2 = \alpha v_x = \alpha v_1$

$\boxed{\dfrac{v_1}{R_1} + C\dfrac{d}{dt}v_1 = i_s}$ $\boxed{-\alpha v_1 + v_2 = 0}$

Note that v_{12} is indeterminant because of the two separate reference points.

12.   (a)   Given the circuit shown below:

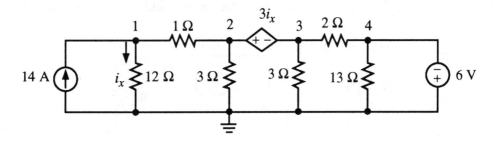

Node 1, KCL: $\frac{1}{12}v_1 + v_1 - v_2 - 14 = 0$

Node 2-3, KCL: $\frac{1}{1}(v_2 - v_1) + \frac{1}{3}v_2 + \frac{1}{3}v_3 + \frac{1}{2}(v_3 - v_4) = 0$

Node 2-3, KVL: $v_2 - v_3 = 3i_x = 3\left(\dfrac{v_1}{12}\right) \Rightarrow \frac{1}{4}v_1 - v_2 + v_3 = 0$

Node 4, KVL: $v_4 = -6$ V

$$\begin{bmatrix} 13/12 & -1 & 0 & 0 \\ -1 & 4/3 & 5/6 & -1/2 \\ 1/4 & -1 & 1 & 0 \\ 0 & 0 & 0 & 1 \end{bmatrix} \cdot \begin{bmatrix} v_1 \\ v_2 \\ v_3 \\ v_4 \end{bmatrix} = \begin{bmatrix} 14 \\ 0 \\ 0 \\ -6 \end{bmatrix} \Rightarrow \boxed{v_1 = 24 \text{ V}} \quad \boxed{v_4 = -6 \text{ V}}$$
$$\boxed{v_2 = 12 \text{ V}}$$
$$\boxed{v_3 = 6 \text{ V}}$$

(b) Referring to part (a), if the two KVL equations are used directly, a 2 X 2 matrix results.

Node 1, KCL: $-14 + \frac{1}{12}v_1 + \frac{1}{1}(v_1 - v_2) = 0$

Node 2-3, KCL: $\frac{1}{1}(v_2 - v_1) + \frac{1}{3}v_2 + \frac{1}{3}\left(v_2 - \frac{1}{4}v_1\right) + \frac{1}{2}\left[v_2 - \frac{1}{4}v_1 - (-6)\right] = 0$

(c) Node 2-3, KCL: $\frac{1}{1}(v_2 - v_1) + \frac{1}{3}v_2 + i_{23} = 0 \Rightarrow i_{23} = 8$ A

$p = (v_2 - v_3)(i_{23}) \Rightarrow \boxed{p = 48 \text{ W}}$

13. Given the circuit shown below:

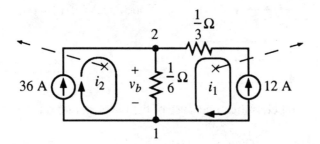

The KCL equations around each mesh yield: $i_1 = -12$ A and $i_2 = 36$ A

Then $i_{12} = i_1 - i_2 \Rightarrow \boxed{i_{12} = -48 \text{ A}}$

$v_b = -v_{12} = -\frac{1}{6}i_{12} \Rightarrow \boxed{v_b = 8 \text{ V}}$

14.   Given the circuit shown below:

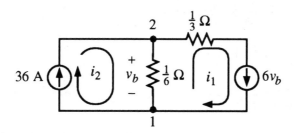

Note that $v_b = \frac{1}{6}i_{21} = \frac{1}{6}(i_2 - i_1)$

Mesh 1, KCL: $i_1 = 6v_b = 1(i_2 - i_1) \Rightarrow i_1 = \frac{1}{2}i_2$

Mesh 2, KCL: $i_2 = 36\ A \Rightarrow i_1 = 18\ A$

$\boxed{i_{12} = i_1 - i_2 = -18\ A}$ and $\boxed{v_b = -\frac{1}{6}i_{12} = 3\ V}$

15.   (a)   Given the circuit shown below:

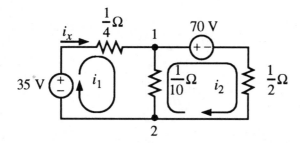

Mesh 1, KVL: $-35 + \frac{1}{4}i_1 + \frac{1}{10}(i_1 - i_2) = 0$

Mesh 2, KCL: $\frac{1}{10}(i_2 - i_1) + 70 + \frac{1}{2}i_2 = 0$

These two equations yield the following matrix equation.

$$\begin{bmatrix} \frac{7}{20} & -\frac{1}{10} \\ -\frac{1}{10} & \frac{3}{5} \end{bmatrix} \cdot \begin{bmatrix} i_1 \\ i_2 \end{bmatrix} = \begin{bmatrix} 35 \\ -70 \end{bmatrix}$$

$\Rightarrow \boxed{i_1 = 70\ V}$ and $\boxed{i_2 = -105\ A}$. The current i_{12} can be found from the

difference of the two mesh currents. $\boxed{i_{12} = i_1 - i_2 = 175\ A}$

    (b)   Refer to part (a) Writing a KVL around the outer loop yields

$-35 + \frac{1}{4}i_1 + 70 + \frac{1}{2}i_2 = 0 \Rightarrow \frac{1}{4}i_1 + \frac{1}{2}i_2 = -35$

Add the two KVL equations found in part (a)

39

Mesh 1, KVL: $-35 + \frac{1}{4}i_1 + \frac{1}{10}(i_1 - i_2) = 0$

Mesh 2, KVL: $\frac{1}{10}(i_2 - i_1) + 70 + \frac{1}{2}i_2 = 0$

$$\frac{1}{4}i_1 + \frac{1}{2}i_2 = -35$$

Notice that this equation is the same as the equation obtained by writing a KVL around the outer loop.

16.   Given the circuit shown below:

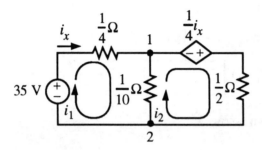

Mesh 1, KVL: $-35 + \frac{1}{4}i_1 + \frac{1}{10}(i_1 - i_2) = 0$

Mesh 2, KVL: $-\frac{1}{4}i_1 + \frac{1}{2}i_2 + \frac{1}{10}(i_2 - i_1) = 0$

$$\begin{bmatrix} \frac{7}{20} & -\frac{1}{10} \\ -\frac{7}{20} & \frac{3}{5} \end{bmatrix} \cdot \begin{bmatrix} i_1 \\ i_2 \end{bmatrix} = \begin{bmatrix} 35 \\ 0 \end{bmatrix}$$

Solving these two equations simultaneously yields

$\boxed{i_1 = 120 \text{ A}}$ and $\boxed{i_2 = 70 \text{ A}}$, and then $\boxed{i_{12} = i_1 - i_2 = 50 \text{ A}}$

17.   Given the circuit shown below:

Mesh 1, KVL: $-35 + \frac{1}{4}i_1 + 10\int_{-\infty}^{t}(i_1 - i_2)d\lambda = 0$

Mesh 2, KVL: $70 + \frac{1}{2}\frac{d}{dt}i_2 + 10\int_{-\infty}^{t}(i_2 - i_1)d\lambda = 0$

$$\begin{bmatrix} \frac{1}{4}+10\int_{-\infty}^{t}d\lambda & -10\int_{-\infty}^{t}d\lambda \\ -10\int_{-\infty}^{t}d\lambda & \frac{1}{2}\frac{d}{dt}+10\int_{-\infty}^{t}d\lambda \end{bmatrix} \cdot \begin{bmatrix} i_1 \\ i_2 \end{bmatrix} = \begin{bmatrix} 35 \\ -70 \end{bmatrix}$$

18. (a) Given the circuit shown below:

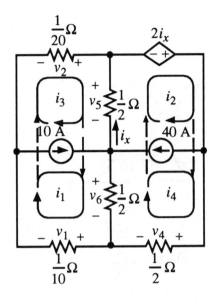

Mesh 1-3, KVL: $\frac{1}{10}i_1 + \frac{1}{20}i_3 + \frac{1}{2}(i_3 - i_2) + \frac{1}{2}(i_1 - i_4) = 0$

Mesh 2-4, KVL: $\frac{1}{2}i_4 + \frac{1}{2}(i_4 - i_1) + \frac{1}{2}(i_2 - i_3) - 2(i_2 - i_3) = 0$

Mesh 1-3, KCL: $i_1 - i_3 = 10$: Mesh 2-4, KCL: $i_2 - i_4 = 40$

Using the two KCL equations, the KVL equations can be written as a function of only i_1 and i_2.

Mesh 1-3, KVL: $\frac{1}{10}i_1 + \frac{1}{20}(i_1 - 10) + \frac{1}{2}(i_1 - i_2 - 10) + \frac{1}{2}(i_1 - i_2 + 40) = 0$

Mesh 2-4, KVL: $\frac{1}{2}(i_2 - 40) + \frac{1}{2}(i_2 - i_1 - 40) + \frac{1}{2}(i_2 - i_1 + 10) - 2(i_2 - i_1 + 10) = 0$

$$\begin{bmatrix} \frac{23}{20} & -1 \\ 1 & -\frac{1}{2} \end{bmatrix} \cdot \begin{bmatrix} i_1 \\ i_2 \end{bmatrix} = \begin{bmatrix} -\frac{29}{2} \\ 55 \end{bmatrix} \Rightarrow \boxed{\begin{array}{l} i_1 = 146.47 \text{ A} \\ i_2 = 182.94 \text{ A} \end{array}}$$

Then the other two mesh currents can be found from the KCL equations.

$\boxed{i_3 = i_1 - 10 = 136.47 \text{ A}}$ and $\boxed{i_4 = i_2 - 40 = 142.94 \text{ A}}$

(b) Refer to part (a)

$$\boxed{v_1 = \tfrac{1}{10}i_1 = 14.647 \text{ V}} \quad : \quad \boxed{v_6 = \tfrac{1}{2}(i_1 - i_4) = 1.765 \text{ V}}$$

The voltage across the 10 A current source is then given by

$$v_{10} = -v_1 - v_6 \implies \boxed{v_{10} = -16.412 \text{ V}}$$

19. Given the circuit shown below:

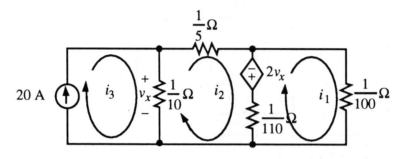

Mesh 3, KCL: $\boxed{i_3 = 20 \text{ A}}$ Note that $v_x = \dfrac{1}{10}(i_3 - i_2) = 2 - \dfrac{1}{10}i_2$

Mesh 1, KVL: $\dfrac{1}{100}i_1 + \dfrac{1}{110}(i_1 - i_2) + 2\left(2 - \dfrac{1}{10}i_2\right) = 0$

Mesh 2, KVL: $\dfrac{1}{10}(i_2 - 20) + \dfrac{1}{5}i_2 - 2\left(2 - \dfrac{1}{10}i_2\right) + \dfrac{1}{110}(i_2 - i_1) = 0$

$$\begin{bmatrix} \frac{21}{1100} & -\frac{23}{110} \\ -\frac{1}{110} & \frac{56}{110} \end{bmatrix} \cdot \begin{bmatrix} i_1 \\ i_2 \end{bmatrix} = \begin{bmatrix} -4 \\ 6 \end{bmatrix} \implies \boxed{\begin{array}{l} i_1 = -100 \text{ A} \\ i_2 = 10 \text{ A} \end{array}}$$

$$p_{dep} = 2v_x(i_1 - i_2)$$

$$p_{dep} = \tfrac{2}{10}(i_3 - i_2)(i_1 - i_2)$$

$$\boxed{p_{dep} = -220 \text{ W}}$$

20. Given the circuit below:

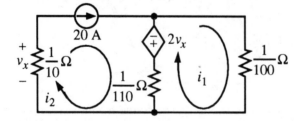

Mesh 2, KCL: $\boxed{i_2 = 20 \text{ A}}$: $v_x = -\dfrac{1}{10}i_2 = -2 \text{ V}$

Mesh 1, KVL: $\dfrac{1}{100}i_1 + \dfrac{1}{110}(i_1 - 20) + 2(-2) = 0 \implies \boxed{i_1 = 219.05 \text{ A}}$

$$p_{dep} = 2v_x(i_1 - i_2) \implies \boxed{p_{dep} = -796.2 \text{ W}}$$

21.   Given the circuit shown below:

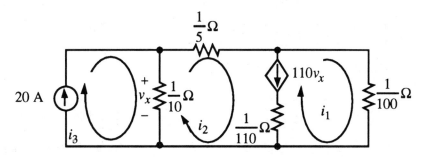

Mesh 1-2, KCL: $i_2 - i_1 = 110v_x = 110\left(\frac{1}{10}\right)(20 - i_2) \Rightarrow i_2 = \frac{55}{3} + \frac{1}{12}i_1$

Mesh 1-2, KVL: $\frac{1}{10}\left[\left(\frac{55}{3} + \frac{1}{12}i_1\right) - 20\right] + \frac{1}{5}\left(\frac{55}{3} + \frac{1}{12}i_1\right) + \frac{1}{100}i_1 = 0$

This KVL equation can be solved for i_1 since it is the only unknown.

$\boxed{i_1 = -100 \text{ A}}$ and therefore $\boxed{i_2 = 10 \text{ A}}$

Now use KVL to determine v_1, the voltage across the dependent source.

$v_1 = \frac{1}{100}i_1 + \frac{1}{110}(i_1 - i_2) \Rightarrow v_1 = -2 \text{ V}$

$p_{dep} = 110v_x(v_1) = 110\left[\frac{1}{10}(20 - 10)\right](-2) \Rightarrow \boxed{p_{dep} = -220 \text{ W}}$

22.   Given the circuit shown below:

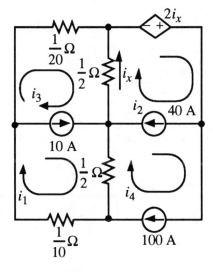

Mesh 4, KCL: $\boxed{i_4 = 100 \text{ A}}$: Mesh 2-4, KCL: $i_2 - i_4 = 40 \Rightarrow \boxed{i_2 = 140 \text{ A}}$

Mesh 1-3, KCL: $i_1 - i_3 = 10 \Rightarrow i_3 = i_1 - 10$

Mesh 1-3, KVL: $\frac{1}{10}i_1 + \frac{1}{20}(i_1 - 10) + \frac{1}{2}(i_1 - 10 - 140) + \frac{1}{2}(i_1 - 100) = 0$

This equation can be solved for i_1 since it is the only unknown.

$$\boxed{i_1 = 109.13\ \text{A}} \quad \Rightarrow \quad \boxed{i_3 = 99.13\ \text{A}}$$

23.   (a)   Given the circuit shown below:

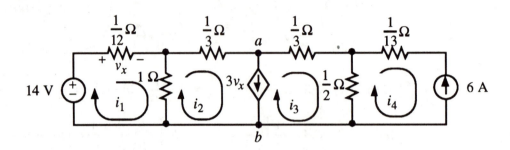

Mesh 1, KVL: $-14 + \dfrac{1}{12}i_1 + \left(i_1 - i_2\right) = 0$

Mesh 2 - 3, KVL: $\left(i_2 - i_1\right) + \dfrac{1}{3}i_2 + \dfrac{1}{3}i_3 + \dfrac{1}{2}\left(i_3 - i_4\right) = 0$

Mesh 2 - 3, KCL: $i_2 - i_3 = 3v_x = 3\left(\dfrac{1}{12}\right)i_1 = \dfrac{1}{4}i_1$: Mesh 4, KCL: $i_4 = -6$

$$\Rightarrow
\begin{bmatrix}
13/12 & -1 & 0 & 0 \\
-1 & 4/3 & 5/6 & -1/2 \\
1/4 & -1 & 1 & 0 \\
0 & 0 & 0 & 1
\end{bmatrix}
\cdot
\begin{bmatrix}
i_1 \\ i_2 \\ i_3 \\ i_4
\end{bmatrix}
=
\begin{bmatrix}
14 \\ 0 \\ 0 \\ -6
\end{bmatrix}$$

(b)   Referring to part (a), if the two KCL equation are used directly, a 2 X 2 matrix results.

Mesh 1, KVL: $-14 + \dfrac{1}{12}i_1 + \dfrac{1}{1}\left(i_1 - i_2\right) = 0$

Mesh 2 - 3, KVL: $\left(i_2 - i_1\right) + \dfrac{1}{3}i_2 + \dfrac{1}{3}\left(i_2 - \dfrac{1}{4}i_1\right) + \dfrac{1}{2}\left[i_2 - \dfrac{1}{4}i_1 - (-6)\right] = 0$

(c)

$$\begin{bmatrix}
\frac{13}{12} & -1 \\
-\frac{29}{24} & \frac{13}{6}
\end{bmatrix}
\cdot
\begin{bmatrix}
i_1 \\ i_2
\end{bmatrix}
=
\begin{bmatrix}
14 \\ -3
\end{bmatrix}
\quad \Rightarrow \quad
\boxed{\begin{matrix} i_1 = 24\ \text{A} \\ i_2 = 12\ \text{A} \end{matrix}}
\quad : \quad \boxed{i_3 = i_2 - \tfrac{1}{4}i_1 = 6\ \text{A}}$$

$$\boxed{i_4 = -6\ \text{A}}$$

Mesh 2, KVL: $\dfrac{1}{1}\left(i_2 - i_1\right) + \dfrac{1}{3}i_2 + v_{ab} = 0 \Rightarrow v_{ab} = 8\ \text{V}$

$p = \left(i_2 - i_3\right)\left(v_{ab}\right) \Rightarrow \boxed{p = 48\ \text{W}}$

24. (a) Number of node voltage equations that are KCL equations
    = Number of nodes -1 -Number of voltage sources
    = 7 - 1 - 5 = 1

    (b) Number of mesh current equations that are KVL equations
    = Number of meshes - Number of current sources
    = 6 - 1 = 5

    (c) Node voltages

25. (a) Number of node voltage equations that are KCL equations
    = Number of nodes - 1 - Number of voltage sources
    = 7 - 1- 1- = 5

    (b) Number of mesh current equations that are KVL equations
    = Number of meshes - Number of current sources
    = 6 - 4 = 2

    (c) Mesh Currents

26. The mesh currents are determined by the current sources.

$$\boxed{i_1 = 2\ A} \qquad \boxed{i_2 = 3\ A}$$

Replace the center 5-A current source with a short circuit and number the nodes as shown below.

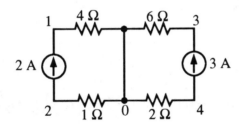

KCL gives

$$-2+\frac{1}{4}v_1 = 0 \ \Rightarrow \ \boxed{v_1 = 8\ V}$$

$$2+\frac{1}{1}v_2 = 0 \ \Rightarrow \ \boxed{v_2 = -2\ V}$$

$$-3+\frac{1}{6}v_3 = 0 \ \Rightarrow \ \boxed{v_3 = 18\ V}$$

$$3+\frac{1}{2}v_4 = 0 \ \Rightarrow \ \boxed{v_4 = -6\ V}$$

$$i_{10} = \frac{1}{4}\left(v_1 - 0\right) = \frac{1}{4}(8) \ \Rightarrow \ \boxed{i_{10} = 2\ A}$$

27.

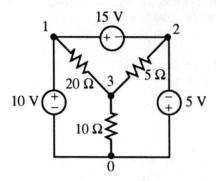

The voltage sources give $\boxed{v_1 = 10\text{ V}}$ and $\boxed{v_2 = -5\text{ V}}$

From KCL applied to a surface enclosing node 3,

$$\frac{1}{20}(v_3 - 10) + \frac{1}{10}v_3 + \frac{1}{5}(v_3 + 5) = 0 \;\Rightarrow\; \boxed{v_3 = -\frac{10}{7}\text{ V}}$$

Replace the 15 V source with an open circuit

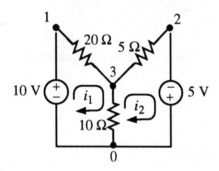

KVL gives $\quad -10 + 20i_1 + 10(i_1 - i_2) = 0 \quad$ and $\quad -5 + 10(i_2 - i_1) + 5i_2 = 0$

$$\begin{bmatrix} 30 & -10 \\ -10 & 15 \end{bmatrix} \cdot \begin{bmatrix} i_1 \\ i_2 \end{bmatrix} = \begin{bmatrix} 10 \\ 5 \end{bmatrix} \;\Rightarrow\; \begin{bmatrix} i_1 \\ i_2 \end{bmatrix} = \begin{bmatrix} \frac{4}{7} \\ \frac{5}{7} \end{bmatrix}$$

$$v_3 = 10(i_1 - i_2) = 10\left[\frac{4}{7} - \frac{5}{7}\right] \;\Rightarrow\; \boxed{v_3 = -\frac{10}{7}\text{ V}}$$

CHAPTER 5

Network Properties

Exercises

1.   (a)

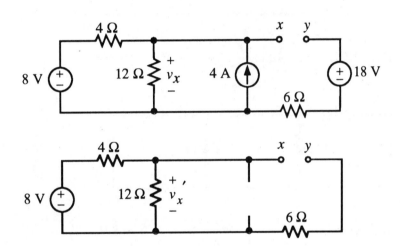

The voltage divider relationship gives $v_x' = \dfrac{12}{4+12} = 6$ V

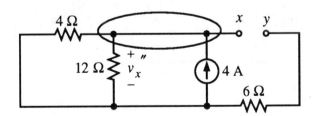

From KCL $\dfrac{1}{4}v_x'' + \dfrac{1}{12}v_x'' - 4 = 0$ $v_x'' = 12$ V

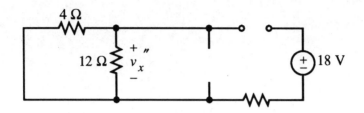

KCL gives $\dfrac{1}{4}v_x''' + \dfrac{1}{12}v_x''' = 0 \qquad v_x''' = 0$

$v_x = v_x' + v_x'' + v_x''' = 6 + 12 + 0 \Rightarrow \boxed{v_x = 18 \text{ V}}$

(b)

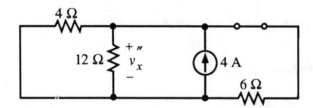

From KCL $\dfrac{1}{4}\left(v_x' - 8\right) + \dfrac{1}{12}v_x' + \dfrac{1}{6}v_x' = 0 \qquad v_x' = 4 \text{ V}$

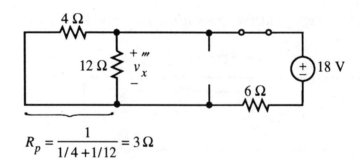

KCL gives $\dfrac{1}{4}v_x'' + \dfrac{1}{12}v_x'' - 4 + \dfrac{1}{6}v_x'' = 0 \qquad v_x'' = 8 \text{ V}$

$R_p = \dfrac{1}{1/4 + 1/12} = 3\,\Omega$

The voltage divider relationship gives $v_x''' = \dfrac{3}{3+6}18 = 6 \text{ V}$

$v_x = v_x' + v_x'' + v_x''' = 4 + 8 + 6 \Rightarrow \boxed{v_x = 18 \text{ V}}$

2.

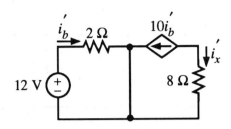

Ohm's law $i_b' = \dfrac{12}{2} = 6$ A

KCL $i_x' = -10i_b' = -60$ A

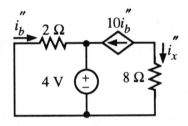

Ohm's law $i_b'' = -\dfrac{4}{2} = -2$ A

KCL $i_x'' = -10i_b'' = -10(-2) = 20$ A

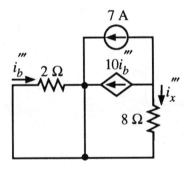

Ohm's law $i_b''' = \dfrac{0}{2} = 0$ A

KCL $7 + 10i_b''' + i_x''' = 0$ $\qquad i_x''' = -7$ A

$i_x = i_x' + i_x'' + i_x'' = -60 + 20 - 7 \Rightarrow \boxed{i_x = -47 \text{ A}}$

3.

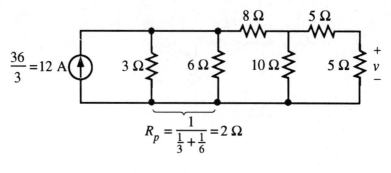

$$R_p = \frac{1}{\frac{1}{3}+\frac{1}{6}} = 2\,\Omega$$

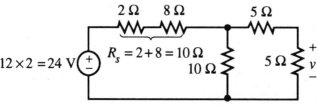

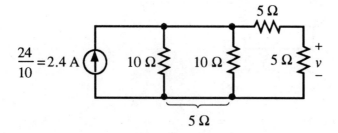

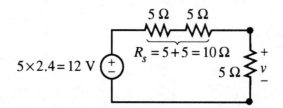

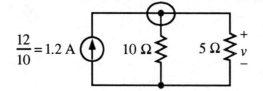

KCL $-1.2 + \frac{1}{10}v + \frac{1}{5}v = 0$

$v = 4$ V

4.

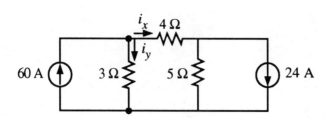

Two source transformations yield

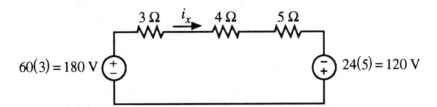

KVL applied to the loop gives

$$-180 + 3i_x + 4i_x + 5i_x - 120 = 0 \implies \boxed{i_x = 25 \text{ A}}$$

Now apply KCL to the top left-hand node of the original circuit.

$$-60 + i_y + i_x = 0$$

$$i_y = 60 - 25 = \boxed{35 \text{ A}}$$

5.   Given the circuit shown below:

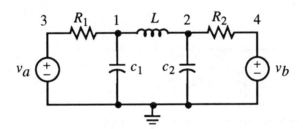

After source transformations

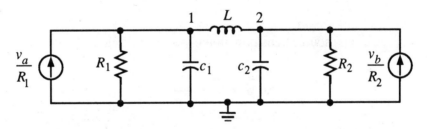

Node 1, KCL: $\quad -\dfrac{1}{R_1}v_a + \dfrac{1}{R_1}v_1 + C_1\dfrac{d}{dt}v_1 + \dfrac{1}{L}\displaystyle\int_{-\infty}^{t}(v_1 - v_2)d\lambda = 0$

Node 2, KCL: $\quad -\dfrac{1}{R_2}v_b + \dfrac{1}{R_2}v_2 + C_2\dfrac{d}{dt}v_2 + \dfrac{1}{L}\displaystyle\int_{-\infty}^{t}(v_2 - v_1)d\lambda = 0$

$$\begin{bmatrix} C_1\dfrac{d}{dt} + \dfrac{1}{R_1} + \dfrac{1}{L}\int_{-\infty}^{t}d\lambda & -\dfrac{1}{L}\int_{-\infty}^{t}d\lambda \\[4mm] -\dfrac{1}{L}\int_{-\infty}^{t}d\lambda & C_2\dfrac{d}{dt} + \dfrac{1}{R_2} + \dfrac{1}{L}\int_{-\infty}^{t}d\lambda \end{bmatrix} \begin{bmatrix} v_1 \\[4mm] v_2 \end{bmatrix} = \begin{bmatrix} \dfrac{1}{R_1}v_a \\[4mm] \dfrac{1}{R_2}v_b \end{bmatrix}$$

6.　　(a)

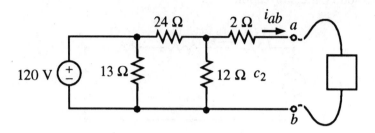

The 120-V source and 13-Ω parallel resistance are equivalent to a 120 V source, as far as v_{ab} and i_{ab} are concerned.

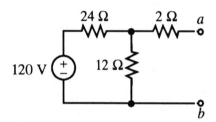

A source transformation give

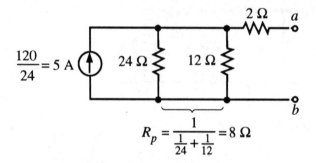

$$R_p = \frac{1}{\frac{1}{24} + \frac{1}{12}} = 8\ \Omega$$

A second source transformation gives

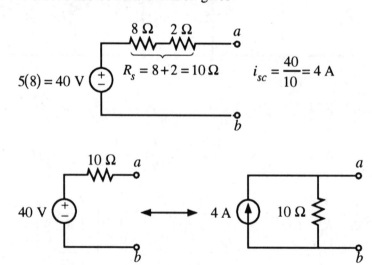

$$R_s = 8 + 2 = 10\ \Omega \qquad i_{sc} = \frac{40}{10} = 4\ \text{A}$$

(b)

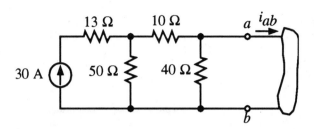

The 30-A source and series 13-Ω resistance are equivalent to a 30-A source, as far as v_{ab} and i_{ab} are concerned.

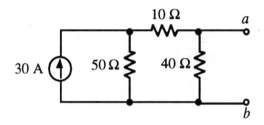

A source transformation gives

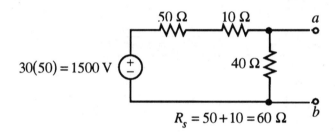

$$30(50) = 1500 \text{ V}$$

$$R_s = 50 + 10 = 60 \ \Omega$$

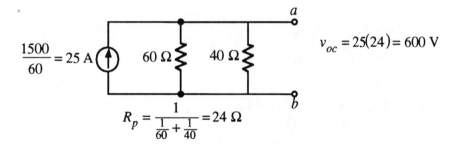

$$\frac{1500}{60} = 25 \text{ A}$$

$$R_p = \frac{1}{\frac{1}{60} + \frac{1}{40}} = 24 \ \Omega$$

$$v_{oc} = 25(24) = 600 \text{ V}$$

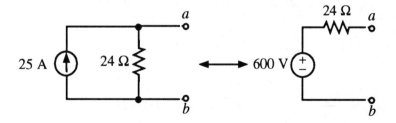

7.

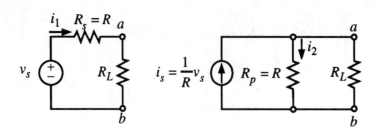

The voltage divider gives

$$i_1 = \frac{1}{R+R_L}v_s = i_2 = \frac{\frac{1}{R}}{\frac{1}{R}+\frac{1}{R_L}}\left(\frac{1}{R}v_s\right)$$

$$\frac{1}{R+R_L} = \frac{R_L}{R}\frac{1}{R+R_L} \Rightarrow \boxed{R_L = R}$$

8.   (a) $R_s = 2+10 = 12,\qquad R_p = \dfrac{1}{\frac{1}{4}+\frac{1}{6}+\frac{1}{12}} = 2\,\Omega$

$$\boxed{R_N = R_{TH} = 2\,\Omega}$$

(b) $R_{TH} = 2 + \dfrac{1}{\frac{1}{0.1}+\frac{1}{4}+\frac{1}{6}} \Rightarrow \boxed{R_{TH} = R_N = 2.096\,\Omega}$

$$\text{KCL}\quad \frac{1}{0.1}\left(v_b - 12\right) + \frac{1}{4}v_b + \frac{1}{6}v_b = 0 \Rightarrow \boxed{v_b = 11.52\ \text{V}}$$

$$v_{oc} = v_b = 11.52\ \text{V}$$

$$i_{sc} = \frac{v_{oc}}{R_{TH}} = \frac{11.52}{2.096} \Rightarrow \boxed{i_{sc} = 5.4962\ \text{A}}$$

(c) $R_{TH} = \dfrac{1}{\frac{1}{0.1}+\frac{1}{6}+\frac{1}{12}} \Rightarrow \boxed{R_{TH} = R_N = 0.0976\,\Omega}$

$$\frac{1}{0.1}\left(v_b - 12\right) + \frac{1}{6}v_b + \frac{1}{12}v_b = 0 \Rightarrow \boxed{v_b = 11.7073\ \text{V}}$$

$$i_{sc} = \frac{v_{bc}}{i_{sc}} = \frac{11.7073}{0.0976} \Rightarrow \boxed{i_{sc} = 120\ \text{A}}$$

(d) $R_{TH} = \dfrac{1}{\frac{1}{0.1}+\frac{1}{4}+\frac{1}{12}} \Rightarrow \boxed{R_{TH} = R_N = 0.0968\,\Omega}$

$$i_{sc} = \frac{12}{0.1} = \boxed{120\ \text{A}} \qquad v_{oc} = R_{TH}i_{sc} = \boxed{11.6129\ \text{V}}$$

Copyright © John Wiley & Sons. All rights reserved.

9.   (a)

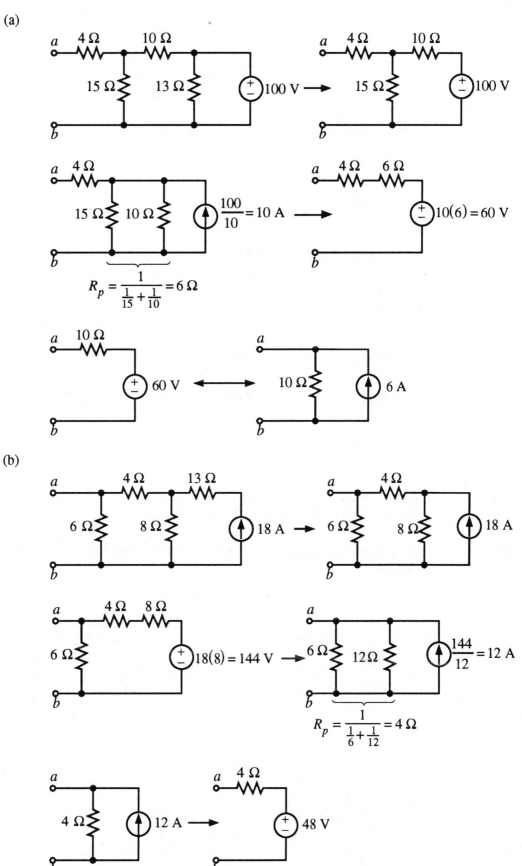

(b)

10.

$$\text{KCL:} \quad -10+\frac{1}{4}v_x+1v_x=0 \implies v_x=\frac{4}{5}=8 \text{ V}$$

$$v_{oc}=v_{ab}=v_x-16(1v_x)=-15v_x-15(8)=-120 \text{ V}$$

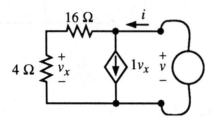

Voltage divider $v_x=\dfrac{4}{4+16}v=\dfrac{1}{5}v$

$$\text{KCL} \quad \frac{1}{20}v+1\left(\frac{1}{5}v\right)-i=0$$

$$i=\frac{5}{20}v=\frac{1}{4}v, \implies R_{in}=\frac{v}{i}=4 \text{ }\Omega.$$

Then $R_{TH}=6+R_{in}=6+4=10 \text{ }\Omega$

$$i_{sc}=\frac{v_{oc}}{R_{TH}}=\frac{-120}{10}=-12 \text{ A}$$

11. R_2 is fixed, so the maximum power is absorbed by R_2 when the current is maximized. This occurs when $R_1=0$.

12.   The source resistance $R_2=5 \text{ }\Omega$ is fixed. The maximum power is absorbed by R_1 when $R_1=R_2=5 \text{ }\Omega$.

13.

$$v_{oc}=v_{ab}\Big|_{i_{ab}=0}=\frac{12}{6+12}36=24 \text{ V}$$

$$R_{TH}=\frac{1}{1/6+1/12}=4 \text{ }\Omega, \qquad R_L=R_{TH}=4 \text{ }\Omega$$

$$i_L=\frac{1}{4+R_L}24=\frac{1}{4+4}24=3 \text{ A} \qquad P_L=i_L^2R_L=3^2(4)=36 \text{ W}$$

14.   From Problem 10, $R_{TH} = 10\,\Omega$

For maximum power delivered to the load, $R_L = R_{TH} = 10\,\Omega$

15.

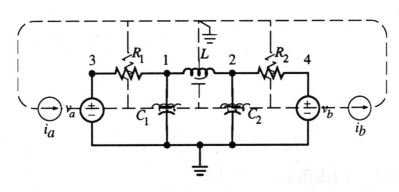

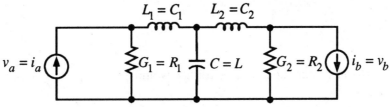

CHAPTER 6

Operational Amplifiers

Exercises

1.   The maximum open circuit voltage would be

$$\max\{v_{oc}\} = V^+ = 12 \text{ V}$$

$$\max\{i_{sc}\} = \frac{1}{R_o}12 = \frac{12}{60} = 0.2 \text{ A} = 200 \text{ mA}$$

2.   (a)   Given the circuit shown below:

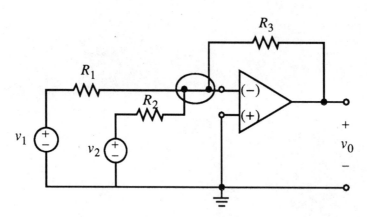

KCL: $-\dfrac{1}{R_1}v_1 - \dfrac{1}{R_2}v_2 - \dfrac{1}{R_3}v_0 = 0 \;\Rightarrow\; \boxed{v_0 = -\dfrac{R_3}{R_1}v_1 - \dfrac{R_3}{R_2}v_2}$

When $R_1 = R_2 = R_3$, $v_0 = -(v_1 + v_2)$

This output is the negative sum of the inputs v_1 and v_2.

(b)   An equivalent circuit for the circuit in Exercise 2a is shown below.

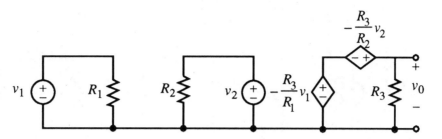

Note that R_3 has no effect since it is in parallel with the voltage source and can therefore be omitted.

3.   (a)   Given the circuit shown below:

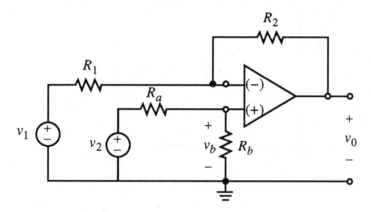

Voltage divider: $v_b = \dfrac{R_b}{R_a + R_b} v_2$

KCL: $\dfrac{1}{R_1}(v_b - v_1) + \dfrac{1}{R_2}(v_b - v_o) = 0 \Rightarrow v_o = \left(\dfrac{R_2}{R_1} + 1\right) v_b - \dfrac{R_2}{R_1} v_1$

$$\boxed{v_o = \dfrac{R_2 + R_1}{R_1}\left(\dfrac{R_b}{R_a + R_b}\right) v_2 - \dfrac{R_2}{R_1} v_1}$$

When $R_1 = R_a$ and $R_2 = R_b$

$$\boxed{v_o = \dfrac{R_2}{R_1}(v_2 - v_1)}$$

(b)   An equivalent circuit for the circuit shown in Exercise 3a is shown below.

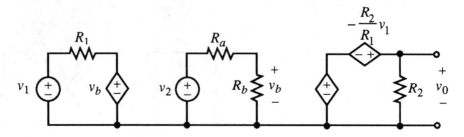

Note that R_2 has no effect on the output and can be removed.

4.   Given the circuit shown below:

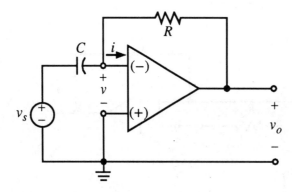

Since $i = 0$, $v = 0$,

KCL: $-C\dfrac{d}{dt}v_s - \dfrac{1}{R}v_o = 0 \Rightarrow \boxed{v_o = -RC\dfrac{d}{dt}v_s}$

5.   (a)   Given the circuit shown below:

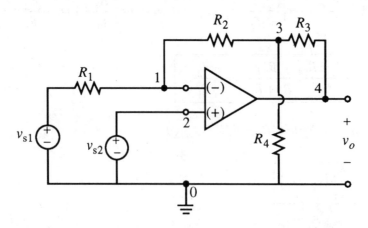

KVL: $v_1 = v_{s2}$

KCL: $\dfrac{1}{R_1}\left(v_{s2} - v_{s1}\right) + \dfrac{1}{R_2}\left(v_{s2} - v_3\right) = 0 \Rightarrow \boxed{v_3 = \dfrac{R_1 + R_2}{R_1}v_{s2} - \dfrac{R_2}{R_1}v_{s1}}$

KCL: $\dfrac{1}{R_2}(v_3 - v_{s2}) + \dfrac{1}{R_4}v_3 + \dfrac{1}{R_3}(v_3 - v_4) = 0$

Rearranging the second equation yields

$$\frac{R_3 R_4 + R_2 R_4 + R_2 R_3}{R_2 R_4}v_3 - \frac{1}{R_2}v_{s2} - \frac{1}{R_3}v_4 = 0$$

$$v_o = v_4 = \frac{R_3 R_4 + R_2 R_4 + R_2 R_3}{R_2 R_4}\left(\frac{R_1 + R_2}{R_1}v_{s2} - \frac{R_2}{R_1}v_{s1}\right) - \frac{R_3}{R_2}v_{s2}$$

$$v_o = \frac{(R_1 + R_2)(R_3 R_4 + R_2 R_4 + R_2 R_3) - R_1 R_3 R_4}{R_1 R_2 R_4}v_{s2} - \frac{R_3 R_4 + R_2 R_4 + R_2 R_3}{R_1 R_4}v_{s1}$$

$$\boxed{v_o = \frac{(R_1 + R_2)(R_3 + R_4) + R_3 R_4}{R_1 R_4}v_{s2} - \frac{R_2(R_3 + R_4) + R_3 R_4}{R_1 R_4}v_{s1}}$$

(b)   Construct an equivalent circuit for the circuit shown in Exercise 5a.

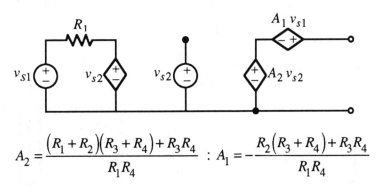

$$A_2 = \frac{(R_1 + R_2)(R_3 + R_4) + R_3 R_4}{R_1 R_4} \quad : \quad A_1 = -\frac{R_2(R_3 + R_4) + R_3 R_4}{R_1 R_4}$$

The load represented by the feedback circuit has been ignored because it has no effect on the output.

6.

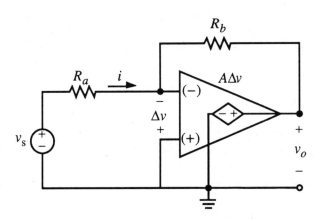

(a)   KCL gives

$$\frac{1}{R_a}\left(-\Delta v - v_s\right) + \frac{1}{R_b}\left(-\Delta v - A\Delta v\right) = 0$$

$$\Delta v = -\frac{R_b}{(1+A)R_a + R_b}v_s$$

$$G = \frac{v_o}{v_s} = \frac{A\Delta v}{v_s} \quad \Rightarrow \quad \boxed{G = -\frac{AR_b}{(1+A)R_a + R_b}}$$

(b) $\quad \lim_{A\to\infty} G = \lim_{A\to\infty}\left[-\frac{AR_b}{(1+A)R_a + R_b}\right] = -\frac{R_b}{R_a}$

Yes, this is the closed-loop gain given by the ideal op-amp model.

(c) $\quad i = \frac{1}{R_a}\left(v_s + \Delta v\right) = \frac{1}{R_a}\left[1 - \frac{R_b}{(1+A)R_a + R_b}\right]v_s$

$$R_i' = \frac{v_s}{i} = R_a\frac{(1+A)R_a + R_b}{(1+A)R_a} = \frac{(1+A)R_a + R_b}{(1+A)}$$

7.

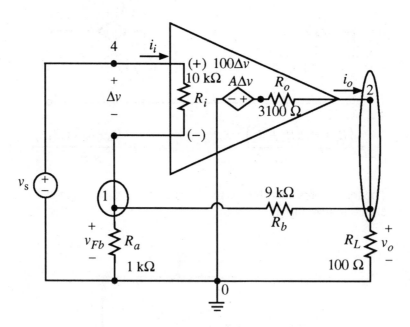

(a) KCL for a surface enclosing node 1 gives

$$\frac{1}{1000}v_1 + \frac{1}{1000}\left(v_1 - v_s\right) + \frac{1}{9000}\left(v_1 - v_2\right) = 0$$

KCL for a surface enclosing node 2 gives

$$\frac{1}{100}v_2 + \frac{1}{9000}\left(v_2 - v_1\right) + \frac{1}{100}\left[v_2 - 100\left(v_s - v_1\right)\right] = 0$$

Operational Amplifiers

Multiply the first KCL equation by 90,000 and the second by 9000:

$$\begin{bmatrix} 109 & -10 \\ 8999 & 181 \end{bmatrix}\begin{bmatrix} v_1 \\ v_2 \end{bmatrix}=\begin{bmatrix} 9v_s \\ 9000v_s \end{bmatrix}$$

The solution is

$$\begin{bmatrix} v_1 \\ v_2 \end{bmatrix}=\begin{bmatrix} 0.83512v_s \\ 8.20285v_s \end{bmatrix} \Rightarrow \text{The closed loop gain is} \boxed{\frac{v_2}{v_s}=8.203}$$

This same result can also be obtained form Eq. (6.39). We can now calculate R_o by use of the short-circuit current and open-circuit voltage. Replace the resistance R_L with a short circuit.

KCL for node 1 gives

$$\frac{1}{1000}v_1+\frac{1}{10000}\left(v_1-v_s\right)+\frac{1}{9000}v_1=0$$

$$v_1=\frac{9}{109}=0.08257v_s$$

The current through the short to ground is

$$i_{sc}=\frac{1}{9000}v_1+\frac{1}{100}\left[100\left(v_s-v_1\right)\right]$$

$$=\frac{1}{9000}\frac{9}{109}v_s+v_s-\frac{9}{109}v_s$$

$$=\frac{900,009}{981,000}v_s=0.9174v_s$$

Now replace R_L with an open circuit. The KCL equation for the surface enclosing node 1 is unchanged from the first KCL equation. The KCL equation for a surface enclosing node 2 becomes

$$\frac{1}{9000}\left(v_2-v_1\right)+\frac{1}{100}\left[v_2-100\left(v_s-v_1\right)\right]=0$$

and we have

$$\begin{bmatrix} 109 & -10 \\ 8999 & 181 \end{bmatrix}\begin{bmatrix} v_1 \\ v_2 \end{bmatrix}=\begin{bmatrix} 9v_s \\ 9000v_s \end{bmatrix}$$

$$v_2=v_{oc}=\frac{900,009}{99,909}v_s=9.008v_s$$

$$R_o'=\frac{v_{oc}}{i_{sc}}=\frac{981,000v_s}{99,909v_s}=9.8189\ \Omega\cong9.82\ \Omega$$

(b) $\quad\beta=\frac{R_a}{R_a+R_b}=\frac{1000}{1000+9000}=\frac{1}{10}$

Copyright © John Wiley & Sons. All rights reserved.

63

$$\frac{v_o}{v_s} = \frac{A}{A\beta + 1} = \frac{100}{\frac{100}{10} + 1} = 9.091$$

$$R_o' = \frac{R_o}{A\beta + 1} = \frac{100}{100\frac{1}{10} + 1} = 9.09\ \Omega$$

The approximation for voltage gain that is about 1% too large. The approximation gives an output resistance that is about 7% less than the actual value. The results are reasonably close even though the loop gain is only 10, and $R_o = 100\ \Omega = R_L$

(c) $$\frac{v_o}{v_s} = \frac{1}{\beta} = 10$$

$$R_o' = \frac{1}{\beta A} R_o = \frac{1}{\frac{1}{10}(100)} 100 = 10\ \Omega$$

The approximation for voltage gain is about 11% too large. The approximation for the output resistance is about 1.8% too low. This approximation is not very good because the loop gain is only 10.

CHAPTER 7

Signal Models

Exercises

1.   (a)

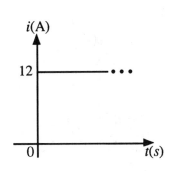

$$i(t) = 12u(t) \text{ A}$$

(b)

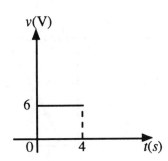

$$v(t) = 6\big[u(t) - u(t-4)\big] \text{ V}$$

(c)

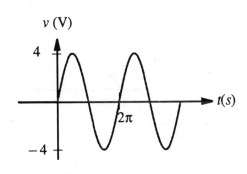

(d)

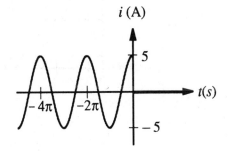

2.   (a) $10u(t-1) - 5u(t-2) - 5u(t-3)$ V

(b) $3t\big[u(t) - u(t-1)\big]$ A

3.   (a) $3u(t-2)$ (b) $2u(2-t)$

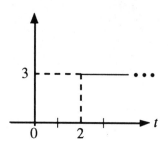   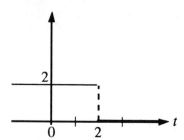

(c) $u(t)+2u(t-1)-u(t-3)$ (d) $tu(t-1)$

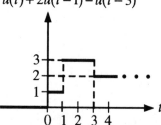   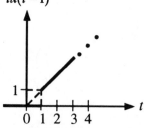

(e) $2(t-1)u(t-1)$ (f) $2u(\sin \pi t)$

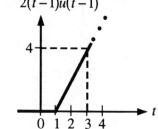   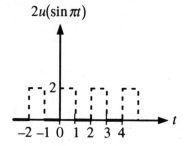

4.   (a) $i = 6u(t)$

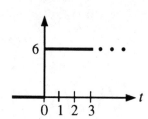

$$v = \begin{cases} \dfrac{1}{2}\displaystyle\int_{-\infty}^{t}0\,d\lambda = 0 & t<0 \\[4mm] \dfrac{1}{2}\left[\displaystyle\int_{-\infty}^{0}0\,d\lambda + \int_{0}^{t}6\,d\lambda\right] = 3t & t>0 \end{cases} = 3tu(t) \ \text{V}$$

(b) $i = 8\big[u(t)-u(t-1)\big]$

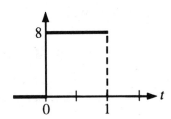 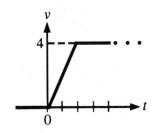

$$v = \begin{cases} \frac{1}{2}\int_{-\infty}^{t}0\,d\lambda = 0 & t < 0 \\ \frac{1}{2}\left[\int_{-\infty}^{0}0\,d\lambda + \int_{0}^{t}8\,d\lambda\right] = 4t & 0 < t < 1 \\ = \frac{1}{2}\left[\int_{-\infty}^{0}0\,d\lambda + \int_{0}^{1}8\,d\lambda + \int_{1}^{t}0\,d\lambda = 4\right] & t > 1 \end{cases} = 4tu(t) - 4(t-1)u(t-1)\ \mathrm{V}$$

5.   (a) $\int_{-\infty}^{\infty}(\cos 2\pi t)\delta(t)dt = \cos 0 = 1$

    (b) $\int_{-\infty}^{\infty}(\cos 2\pi t)\delta(t-1)dt = \cos 2\pi = 1$

    (c) $\int_{-\infty}^{\infty}(\cos 2\pi\lambda)\delta(t-\lambda)d\lambda = \cos 2\pi t$

    (d) $\int_{-\infty}^{\infty}t^3\delta(t-2)dt = 2^3 = 8$

6.   (a) $10e^{-5t}\big|_{t=0.2} = 10e^{-1} = 3.679$

    (b) $8e^{-20t}\big|_{t=0.2} = 8e^{-4} = 0.1465$

    (c) $2e^{10t}\big|_{t=0.2} = 14.778$

    (d) $\frac{d}{dt}20e^{-4t}\big|_{t=0.2} = -80e^{-4t}\big|_{t=0.2} = -80e^{-0.8} = -35.95$

7. $i = C\frac{d}{dt}v = 5\times10^{-6}\frac{d}{dt}40e^{-2t}$

     $= (5\times10^{-6})(-80e^{-2t}) = -0.0004e^{-2t}$

     $= -0.4e^{-2t}\ \mathrm{mA}$

8.   (a) $e^{-0.5t}\big|_{t=1} = e^{-0.5} = 0.6065$

    (b) $\cos 0.5t\big|_{t=1} = \cos 0.5 = 0.8776$

(c) $\quad \cos[0.1t + (\pi/6)]\big|_{t=1} = \cos[0.1 + (\pi/6)] = \cos 0.6236 = 0.8118$

(d) $\quad \cos[0.1t - 30°]\big|_{t=1} = \cos[0.1 - (\pi/6)] = \cos(-0.4236) = 0.9116$

(e) $\quad e^{j0.5t}\big|_{t=1} = \cos 0.5t + j\sin 0.5t\big|_{t=1} = \cos 0.5 + j\sin 0.5 = 0.8776 + j0.4794$

(f) $\quad e^{j[0.5t + (\pi/6)]}\big|_{t=1} = \cos[0.5 + (\pi/6)] + j\sin[0.5 + (\pi/6)]$

$$= \cos 1.0236 + j\sin 1.0236 = 0.5203 + j0.8540$$

(g) $\quad e^{(-2 + j0.5)t}\big|_{t=1} = e^{-2}(\cos 0.5 + j\sin 0.5) = 0.1188 + j0.06488$

9. $\quad v(t) = e^{4t}\cos 3t$

$$= \Re_e\{e^{4t}e^{j3t}\} = \Re_e\{e^{(4+j3)t}\}$$

$$\frac{d}{dt}v(t) = \frac{d}{dt}\Re_e\{e^{(4+j3)t}\} = \Re_e\left\{\frac{d}{dt}e^{(4+j3)t}\right\}$$

$$= \Re_e\{(4+j3)e^{(4+j3)t}\}$$

$$= \Re_e\{5e^{j36.87°}e^{(4+j3)t}\}$$

$$= \Re_e\{5e^{4t}e^{j(3t+36.87°)}\}$$

$$= 5e^{4t}\cos(3t + 36.87°)$$

10. (a) $\quad I_{dc} = \lim_{T\to\infty}\frac{1}{T}\int_{-T/2}^{T/2}i_a dt = \lim_{T\to\infty}\frac{1}{T}\int_{-T/2}^{T/2}5dt = \lim_{T\to\infty}\frac{1}{T}5T = 5\,\text{A}$

(b) $\quad v_{ab}$ is preiodic with period $T = 2\pi/w$, so the average over the infinite interval is the same as the average over 1 period.

$$V_{dc} = \frac{\omega}{2\pi}\int_{-\pi/2}^{\pi/2}[4\cos\omega t + 3\cos\omega t]dt = \frac{\omega}{2\pi}\int_{-\pi/\omega}^{\pi/\omega}7\cos\omega t dt = \frac{\omega}{2\pi}\frac{7}{\omega}\sin\omega t\Big|_{\frac{-\pi}{\omega}}^{\frac{\pi}{\omega}} = 0$$

This result is easy to visualize from the plot of a cosine.

(c) $\quad I_{dc} = \lim_{T\to\infty}\frac{1}{T}\int_{-T/2}^{T/2}4 + 3\cos 2t dt$

$$= \lim_{T\to\infty}\frac{1}{T}\int_{-T/2}^{T/2}4dt + \lim_{T\to\infty}\frac{1}{T}\int_{-T/2}^{T/2}3\cos 2t dt = 4 + 0 = 4\,\text{A}$$

(d) $\quad V_{dc} = \lim_{T \to \infty} \frac{1}{T} \int_{-T/2}^{T/2} 8\cos^2 \omega t\, dt + \lim_{T \to \infty} \frac{1}{T} \int_{-T/2}^{T/2} [4 + 4\cos 2\omega t]\, dt$

$\quad\quad = \lim_{T \to \infty} \frac{1}{T} \int_{-T/2}^{T/2} 4\, dt + \lim_{T \to \infty} \frac{1}{T} \int_{-T/2}^{T/2} 4\cos 2\omega t\, dt = 4 + 0 = 4 \text{ V}$

11. (a) $\quad I_{rms} = \sqrt{\lim_{T \to \infty} \frac{1}{T} \int_{-T/2}^{T/2} i_a^2\, dt} = \sqrt{\lim_{T \to \infty} \frac{1}{T} \int_{-T/2}^{T/2} 5^2\, dt} = \sqrt{25} = 5 \text{ A}$

(b) $\quad V_{rms} = \sqrt{\lim_{T \to \infty} \frac{1}{T} \int_{-T/2}^{T/2} [3 + 4\cos 2t]^2\, dt}$

$\quad\quad = \sqrt{\lim_{T \to \infty} \frac{1}{T} \int_{-T/2}^{T/2} [9 + 24\cos 2t + 16\cos^2 2t]\, dt}$

$\quad\quad = \sqrt{\lim_{T \to \infty} \frac{1}{T} \int_{-T/2}^{T/2} [9 + 24\cos 2t + 8 + 8\cos 4t]\, dt}$

$\quad\quad = \sqrt{9 + 0 + 8 + 0} = \sqrt{17} \text{ V}$

12. (a) $\quad v_a = 7\cos(\omega t) \Rightarrow \boxed{V_{rms} = \frac{7}{\sqrt{2}} = 4.95 \text{ V}_{rms}}$

(b) $\quad v_b = 4\cos(\omega t) + 3\sin(\omega t) \Rightarrow \boxed{V_{rms} = \sqrt{\frac{4^2}{2} + \frac{3^2}{2}} = \frac{5}{\sqrt{2}} \text{ V}_{rms}}$

(c) $\quad i_c = 4\cos(5t) + 4\cos(10t) \Rightarrow \boxed{I_{rms} = \sqrt{\frac{4^2}{2} + \frac{4^2}{2}} = 4 \text{ A}_{rms}}$

(d) $\quad i_d = 4 + 3\cos(2t) \Rightarrow \boxed{I_{rms} = \sqrt{4^2 + \frac{3^2}{2}} = 4.53 \text{ A}_{rms}}$

(e) $\quad i_e = 2\cos(2t) + 2\sin(2t + 45°) = 2\cos(2t) + \frac{2}{\sqrt{2}}\sin(2t) + \frac{2}{\sqrt{2}}\cos(2t)$

(f) $\quad i_e = \sqrt{(2 + \sqrt{2})^2 + (\sqrt{2})^2}\, \cos\left[2t - \arctan\left(\frac{\sqrt{2}}{2 + \sqrt{2}}\right)\right]$

(g) $\quad i_e = 3.70\cos(2t - 0.393) \Rightarrow \boxed{I_{rms} = \frac{3.70}{\sqrt{2}} = 2.61 \text{ A}_{rms}}$

CHAPTER 8

First-Order Circuits

Exercises

1.   (a) $\dfrac{d}{dt}v + 6v = 0$, $v(0^+) = 12$

has the characteristic equation

$$s + 6 = 0 \Rightarrow \boxed{s = -6}$$

$$v = Ae^{-6t}$$

$$v(0^+) = 12 = Ae^{-6(0)} \Rightarrow \boxed{A = 12}$$

$$\boxed{v = 12e^{-6t} \text{ V} \qquad t > 0}$$

    (b) $3\dfrac{d}{dt}v + 6v = 0$, $v(0^+) = 12$

has the characteristic equation

$$s + 2 = 0 \Rightarrow \boxed{s = -2}$$

$$v = Ae^{-2t}$$

$$v(0^+) = 12 = Ae^{-2(0)} \Rightarrow \boxed{A = 12}$$

$$\boxed{v = 12e^{-2t} \text{ V} \qquad t > 0}$$

    (c) $i + 4\displaystyle\int_{-\infty}^{t} i\,d\lambda = 0$ $i(0^+) = 7$

Differentiation yields

$$\frac{d}{dt}i + 4i = 0$$

This has the characteristic equation

$$s + 4 = 0 \;\Rightarrow\; s = -4$$

$$i = Ae^{-4t}$$

$$i(0^+) = 7 \;\Rightarrow\; A = 7$$

$$\boxed{i = 7e^{-4t} \text{ A} \quad t > 0}$$

2.

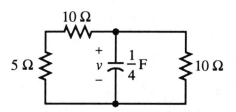

Application of KCL to the top node yields

$$\frac{1}{4}\frac{d}{dt}v + \frac{1}{15}v + \frac{1}{10}v = 0$$

which reduces to

$$\frac{d}{dt}v + \frac{2}{3}v = 0$$

The characteristic equation

$$s + \frac{2}{3} = 0 \;\Rightarrow\; s = -\frac{2}{3} \;\Rightarrow\; \boxed{v = Ae^{-(2/3)t}}$$

$$v(0^+) = 10 = Ae^{-(2/3)(0)} \;\Rightarrow\; \boxed{A = 10}$$

$$\boxed{v = 10e^{-(2/3)t} \text{ V} \qquad t > 0}$$

3.

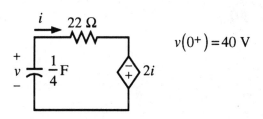

$$v(0^+) = 40 \text{ V}$$

KVL at $t = 0^+$ is

$$-40 + 22i(0^+) - 2i(0^+) = 0$$

$$i(0^+) = 2 \text{ A}$$

KVL

$$4\int_{-\infty}^{t} i d\lambda + 22i - 2i = 0$$

Differentiation and division by 20 yields

$$\frac{d}{dt}i + \frac{1}{5}i = 0$$

The characteristic equation is

$$s + \frac{1}{5} = 0 \implies s = -\frac{1}{5} \implies \boxed{i = Ae^{-t/5}}$$

$$i(0^+) = 2 = Ae^{-0/5} \implies A = 2$$

$$\boxed{i = 2e^{-t/5} \text{ A} \qquad t > 0}$$

4.

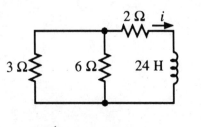

$$v_c(0^+) = 100 \text{ V}$$

$$R = 4 + \frac{1}{\frac{1}{15} + \frac{1}{10}} = 10 \text{ } \Omega$$

$$i_c(0^+) = \frac{1}{10} v_c(0^+) = \frac{100}{10} = 10 \text{ A}$$

$$i(0^+) = \frac{\frac{1}{10}}{\frac{1}{15} + \frac{1}{10}} 10 = 6 \text{ A}$$

$$\tau = RC = 10 \left(\frac{1}{20} \right) = \frac{1}{2}$$

$$i = i(0^+) e^{-t/\tau} = 6e^{-2t} \text{ A} \qquad t > 0$$

5.

$$w = \int_0^\infty p \, dt = \int_0^\infty RI_0^2 e^{-2tR/L} dt$$

$$= RI_0^2 \left(-\frac{L}{2R} \right) e^{-2tR/L} \Big|_0^\infty = \frac{1}{2} LI_0^2$$

6.

$$i(0^+) = 10 \text{ A}$$

$$R_p = \frac{1}{\frac{1}{3} + \frac{1}{6}} = 2 \text{ } \Omega$$

KVL gives

$$24\frac{d}{dt}i + 2i + 2i = 0$$

$$\frac{d}{dt}i + \frac{1}{6}i = 0$$

The characteristic equation is

$$s + \frac{1}{6} = 0 \;\Rightarrow\; s = -\frac{1}{6} \;\Rightarrow\; \boxed{i = Ae^{-t/6}}$$

$$i(0^+) = 10 = Ae^{-0/6} \;\Rightarrow\; \boxed{A = 10}$$

$$\boxed{i = 10e^{-t/6}\ \text{A} \qquad t > 0}$$

7.

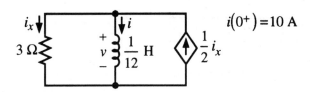

$i(0^+) = 10$ A

The KCL equation at $t = 0^+$ is

$$\frac{1}{3}v(0^+) + 10 - \frac{1}{2}\left[\frac{1}{3}v(0^+)\right] = 0$$

$$v(0^+) = -60\ \text{V}$$

The KCL equation for the top node is

$$\frac{1}{3}v + 12\int_{-\infty}^{t} v\,d\lambda - \frac{1}{2}\left[\frac{1}{3}v\right] = 0$$

Differentiate and multiply by 6.

$$\frac{d}{dt}v + 72v = 0$$

The characteristic equation is

$$s + 72 = 0 \;\Rightarrow\; s = -72 \;\Rightarrow\; v = Ae^{-72t}\ \text{V} \qquad t > 0$$

$$v(0^+) = -60 = Ae^{-72(0)} \;\Rightarrow\; A = -60$$

$$\boxed{v = -60e^{-72t}\ \text{V} \qquad t > 0}$$

8.

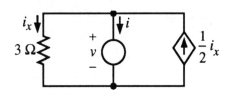

KCL

$$\frac{1}{3}v + i - \frac{1}{2}\left(\frac{1}{3}v\right) = 0$$

$$i = -\frac{1}{6}v$$

$$R = -\frac{v}{i} = 6$$

$$\tau = \frac{L}{R} = \frac{1/12}{6} = \frac{1}{72}\,s$$

(a) $i(0^+) = 10$ A is given

$$\boxed{i = i(0^+)e^{-t/\tau} = 10e^{-72t} \text{ A} \qquad t > 0}$$

(b) $v(0^+) = -Ri(0^+) = -6(10) = -60$ V

$$\boxed{v = v(0^+)e^{-t/\tau} = -60e^{-72t} \text{ V} \qquad t > 0}$$

9.

(a) $i_s = 10u(t)$ A

$$v(0^+) = v(0^-) = 0 = i_s = 0 \qquad t < 0$$

KCL

$$-i_s + \frac{1}{1}v + \frac{1}{4}\frac{d}{dt}v = 0$$

$$\frac{d}{dt}v + 4v = 4i_s$$

$$i_s = 10 \text{ A} \qquad t > 0$$

$$0 + 4v_p = 4(10) \;\Rightarrow\; v_p = 10 \text{ V} \qquad t > 0$$

$$s + 4 = 0 \;\Rightarrow\; s = -4 \;\Rightarrow\; v_n = Ae^{-4t}$$

$$v(0^+) = 0 = 10 + Ae^{-4(0)} \;\Rightarrow\; A = -10$$

$$v = 10 - 10e^{-4t} \text{ V} \qquad t > 0$$

$$\boxed{v = 10\left(1 - e^{-4t}\right)u(t) \text{ V}}$$

(b) $\quad v(0^+) = v(0^-) = 0 = i_s = 0 \qquad t < 0$

$i_s = 10e^{-2t}$ A $\qquad t > 0$

$-2v_p + 4v_p = 4(10e^{-2t}) \implies v_p = 20e^{-2t}$ V $\qquad t > 0$

From part (a) $v_n = Ae^{-4t}$

$v = v_p + v_n = 20e^{-2t} + Ae^{-4t}$

$v(0^+) = 0 = 20e^0 + Ae^0 \implies A = -20$

$$\boxed{v = 20(e^{-2t} - e^{-4t})u(t) \text{ V}}$$

(c) $\quad v(0^+) = v(0^-) = 0 = i_s = 0 \qquad t < 0$

$i_s = 10e^{-4t}$ A $\qquad t > 0$

$s_p = -4$ is the root of the characteristic equation, so

$$\frac{d}{dt}v + 4v = 4(10e^{-4t})$$

has the particular response

$v_p = 40te^{-4t}$

We know from part (a) that

$v_n = Ae^{-4t}$

$v = v_p + v_n = 40te^{-4t} + Ae^{-4t}$

$v(0^+) = 0 = 40(0)e^0 + Ae^0 \implies A = 0$

$$\boxed{v = 40te^{-4t}u(t) \text{ V}}$$

10.

$$v_s = \begin{cases} 50 \text{ V} & t < 0 \\ 150 \text{ V} & t > 0 \end{cases}$$

(a)

From KVL

$-50 + 20(0) + v_p = 0$

$v_p = 50$ V $\qquad t < 0$

$$v(0^-) = v_p(0^-) = 50 \text{ V}$$

$$i(0^-) = i_p(0^-) = 0 \text{ A}$$

(b) $v(0^+) = v(0^-) = 50 \text{ V}$

(c)   KVL for $t = 0^+$ gives

$$-150 + 20i(0^+) + 50 = 0$$

$$i(0^+) = 5 \text{ A}$$

(d) $-v_s + 20i + 40\int_{-\infty}^{t} i \, d\lambda = 0$

Differentiate and divide by 20

$$\frac{d}{dt}i + 2i = \frac{1}{20}\frac{d}{dt}v_s = 0$$

$$(0)i_p + 2i_p = 0$$

$$i_p = 0 \text{ A} \qquad t > 0$$

$$s + 2 = 0 \ \Rightarrow\ s = -2 \ \Rightarrow\ i_n = Ae^{-2t}$$

$$i = i_p + i_n = 0 + Ae^{-2t}$$

$$i(0^+) = 5 = Ae^0 \ \Rightarrow\ A = 5$$

$$\boxed{i = 5e^{-2t} \text{ A} \qquad t > 0}$$

11. $\tau = RC = 20(0.025) = 0.5 \text{ s}$

Replace the capacitance by an open circuit. KVL at $t = 0^-$ and continuity of capacitance voltage yields

$$v(0^+) = v(0^-) = 50 \text{ V}$$

At $t = 0^+$, KVL gives

$$-150 + 20i(0^+) + 50 = 0$$

$$i(0^+) = 5 \text{ A}$$

Replace the capacitance with an open circuit. For $t > 0$

$$i = i(\infty) + \left[i(0^+) - i(\infty)\right]e^{-t/\tau}$$

$$= 0 + [5 - 0]e^{-2t}$$

$$= 5e^{-2t} \text{ A} \qquad t > 0$$

12.

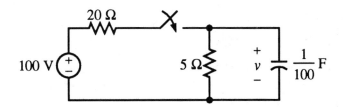

The switch is open for $t < 0$, so

$$v(0^+) = v(0^-) = 0 \text{ V}$$

With the switch closed, replace the capacitance with an open circuit

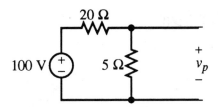

The voltage divider relation gives

$$v(\infty) = v_p = \frac{5}{5+20}100 = 20 \text{ V} \qquad t > 0$$

Set the voltage source to zero to calculate

$$R_{TH} = \frac{1}{\frac{1}{5} + \frac{1}{20}} = 4 \ \Omega$$

$$\tau = RC = 4\left(\frac{1}{100}\right) = \frac{1}{25} \text{ s}$$

$$v = v(\infty) + \left[v(0^+) - v(\infty)\right]e^{-t/\tau}$$

$$= 20 + [0 - 20]e^{-25t}$$

$$= 20\left[1 - e^{-25t}\right] \text{V} \qquad t > 0$$

13.

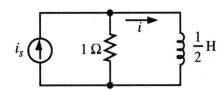

KVL gives

$$\frac{1}{2}\frac{d}{dt}i + 1(i - i_s) = 0$$

and

Chapter 8

$$\frac{d}{dt}i + 2i = 2i_s$$

The characteristic equation is

$$s + 2 = 0 \Rightarrow s = -2 \Rightarrow i_n = Ae^{-2t}$$

(a) $i_s = 2u(t)$

The source is zero for $t < 0$, so

$$i(0^+) = i(0^-) = 0$$
$$i_s = 2 \text{ A} \qquad t > 0, \text{ so}$$

$$0i_p + 2i_p = 2(2) \Rightarrow i_p = 2 \text{ A} \qquad t > 0$$
$$i = i_p + i_n = 2 + Ae^{-2t}$$
$$i(0^+) = 0 = 2 + Ae^0 \Rightarrow A = -2$$

$$\boxed{i = 2(1 - e^{-2t}) \text{ A} \qquad t > 0}$$

(b) $i_s = [2 + 2u(t)] \text{ A}$

$$i_s = 2 \text{ A} \qquad t < 0$$
$$0i_p + 2i_p = 2(2) \Rightarrow i_p = 2 \text{ A} \qquad t < 0$$
$$i(0^+) = i(0^-) = i_p(0^-) = 2 \text{ A}$$
$$i_s = 4 \text{ A} \qquad t > 0$$
$$0i_p + 2i_p = 2(4) \Rightarrow i_p = 4 \text{ A} \qquad t > 0$$
$$i = i_p + i_n = 4 + Ae^{-2t}$$
$$i(0^+) = 2 = 4 + Ae^0 \Rightarrow A = -2$$

$$\boxed{i = 4 - 2e^{-2t} \text{ A} \qquad t > 0}$$

(c) $i_s = 2 \text{ A} \qquad t < 0 \text{ and } 2e^{-6t} \qquad t > 0$

$$0i_p + 2i_p = 2(2) \Rightarrow i_p = 2 \text{ A} \qquad t < 0$$
$$i(0^+) = i(0^-) = i_p(0^-) = 2 \text{ A}$$
$$-6i_p + 2i_p = 2(2e^{-6t}) \Rightarrow i_p = -e^{-6t} \text{ A} \qquad t > 0$$
$$i = i_p + i_n = -e^{-6t} + Ae^{-2t}$$
$$i(0^+) = 2 = -e^0 + Ae^0 \Rightarrow A = 3$$

$$\boxed{i = 3e^{-2t} - e^{-6t} \text{ A} \qquad t > 0}$$

78

14.

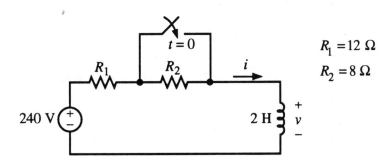

$R_1 = 12 \ \Omega$

$R_2 = 8 \ \Omega$

Replace the inductance with a short circuit to calculate the i_p.
KVL gives

$$-240 + 12i_p + 8i_p + 0 = 0 \implies i_p = 12 \ \text{A} \qquad t < 0$$

KVL for $t > 0$ yields

$$-240 + 12i_p + 0 = 0 \implies i_p = 20 \ \text{A} \qquad t > 0$$

For $t > 0$

$$-240 + 12i + 2\frac{d}{dt}i = 0$$

The characteristic equation is

$$s + 6 = 0 \implies s = -6 \implies i_n = Ae^{-6t} \ \text{A}$$

$$i_p + i_n = 20 + Ae^{-6t}$$

$$i(0^+) = i(0^-) = i_p(0^-) = 12 \ \text{A}$$

$$i(0^+) = 12 = 20 + Ae^0 \implies A = -8$$

$$\boxed{i = 20 - 8e^{-6t} \qquad t > 0}$$

15.

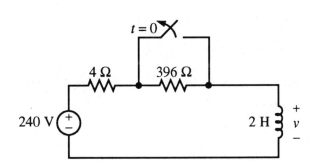

Replace the inductance with a short circuit to find i_p for $t < 0$. KVL gives

$$-240 + 4i_p + 0 = 0 \implies i_p = 60 \ \text{A}$$

$$i(0^+) = i(0^-) = i_p(0^-) = 60 \ \text{A}$$

KVL for $t = 0^+$ i_s

$$-240+4i(0^+)+396i(0^+)+v(0^+)=0$$

$$v(0^+)=240-400(60)=-23,760 \text{ A}$$

If we replace the inductance by a short circuit, we see that

$$v(\infty)=v_p=0 \qquad t>0$$

$$R_{TH}=4+396=400 \text{ }\Omega$$

$$\tau=\frac{L}{R}=\frac{2}{400}=\frac{1}{200}$$

$$v=v(\infty)+\left[v(0^+)-v(\infty)\right]e^{-t/\tau}$$

$$=-23,760e^{-200t} \text{ V} \qquad t>0$$

16.

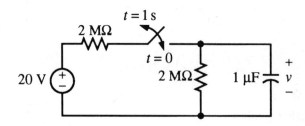

$$v(0^+)=v(0^-)=0 \text{ V}$$

For $0<t<1$s , KCL gives

$$\frac{1}{2\times10^6}(v-20)+\frac{1}{2\times10^6}v+1\times10^{-6}\frac{d}{dt}v=0$$

$$\frac{d}{dt}v+v=10 \quad 0<t<1\text{s}$$

$$0v_p+v_p=10 \Rightarrow v_p=10 \text{ V} \quad 0<t<1\text{s}$$

$$s+1=0 \Rightarrow s=-1 \Rightarrow v_n=Ae^{-t} \quad 0<t<1\text{s}$$

$$v=v_p+v_n=10+Ae^{-t}$$

$$v(0^+)=0=10+Ae^{-0} \Rightarrow A=-10$$

$$\boxed{v=10(1-e^{-t}) \text{ V} \quad 0<t<1\text{s}}$$

$$v(1^-)=10(1-e^{-1})=6.321$$

$$v(1^+)=v(1^-)=10(1-e^{-1})=6.321 \text{ V}$$

For t > 1, KCL gives

$$\frac{1}{2\times10^6}v+1\times10^{-6}\frac{d}{dt}v=0$$

$$\frac{d}{dt}v+\frac{1}{2}v=0 \Rightarrow s+\frac{1}{2}=0 \Rightarrow s=-\frac{1}{2}$$

$$v_n = Ae^{-\frac{1}{2}t} \qquad t>1\,s$$

$$v = v_n + v_p = Ae^{-\frac{1}{2}t}$$

$$v(1^+) = 10(1-e^{-1}) = Ae^{-\frac{1}{2}} \;\Rightarrow\; A = 10(1-e^{-1})e^{\frac{1}{2}}$$

$$v = 10(1-e^{-1})e^{\frac{1}{2}}e^{-\frac{t}{2}} = 10(1-e^{-1})e^{-(t-1)/2}\text{ V} \qquad t>1$$

$$\boxed{v = 6.321e^{-(t-1)/2}\text{ V} \qquad t>1}$$

17.

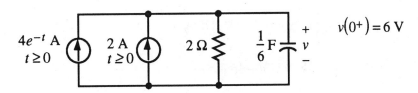

Application of KCL gives

$$-4e^{-t} - 2 + \frac{1}{2}v + \frac{1}{6}\frac{d}{dt}v = 0$$

which we can rewrite as

$$\frac{d}{dt}v + 3v = 12 + 24e^{-t}$$

$$s + 3 = 0 \;\Rightarrow\; s = -3 \;\Rightarrow\; v_n = Ae^{-3t}$$

$$(0)v_{p1} + 3v_{p1} = 12 \;\Rightarrow\; v_{p1} = 4\text{ V} \qquad t>0$$

$$(-1)v_{p2} + 3v_{p2} = 24e^{-t} \;\Rightarrow\; v_{p2} = 12e^{-t} \qquad t>0$$

$$v = v_{p1} + v_{p2} + v_n = 4 + 12e^{-t} + Ae^{-3t}$$

$$v(0^+) = 6 = 4 + 12e^0 + Ae^0 \;\Rightarrow\; A = -10$$

$$\boxed{v = 4 + 12e^{-t} - 10e^{-3t}\text{ V} \qquad t>0}$$

18.

$$v_s = \begin{cases} 20\text{ V} & t<0 \\ 50\cos 3t\text{ V} & t>0 \end{cases}$$

$$\frac{1}{1000}(v - v_s) + 250\times10^{-6}\frac{d}{dt}v = 0$$

which reduces to

$$\left(\frac{d}{dt} + 4\right)v = 4v_s$$

$$(0+4)v_p = 4(20) \;\Rightarrow\; v_p = 20\text{ V} \qquad t>0$$

81

KCL gives

$$v(0^+) = v(0^-) = v_p(0^-) = 20 \text{ V}$$

$$s + 4 = 0 \implies v_n = Ae^{-4t} \text{ V}$$

For $t > 0$

$$(j3 + 4)v_p = 4(50e^{j3t})$$

$$v_p = \frac{200}{4 + j3}e^{j3t} = 40e^{j(3t - 36.87°)}$$

$$v_p = \Re e\{v_p\} = 40\cos(3t - 36.87°) \text{ V}$$

$$v = v_n + v_p = 40\cos(3t - 36.87°) + Ae^{-4t}$$

$$v(0^+) = 20 = 40\cos(0 - 36.87°) + Ae^0$$

$$A = -12$$

$$\boxed{v = 40\cos(3t - 36.87°) - 12e^{-4t} \text{ V} \qquad t > 0}$$

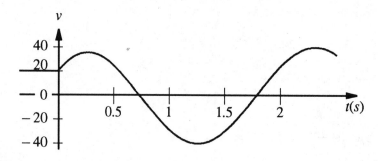

19.

$$v_s = \begin{cases} 50\cos 3t \text{ V} & t < 0 \\ 20 \text{ V} & t > 0 \end{cases}$$

The KCL equation for the top right-hand node reduces to (see Exercise 18)

$$\left(\frac{d}{dt} + 4\right) = 4v_s$$

For $t < 0$,

$$(j3 + 4)v_p = 4(50e^{j3t})$$

$$v_p = \frac{200}{4 + j3}e^{j3t} = 40e^{j(3t - 36.87°)}$$

$$v_p = \Re\mathbf{e}\{\mathbf{v}_p\} = 40\cos(3t - 36.87°)\text{ V} \qquad t < 0$$

$$v(0^+) = v(0^-) = v_p(0^-) = 40\cos(0 - 36.87°)$$

$$= 32\text{ V}$$

$$s + 4 = 0 \implies v_n = Ae^{-4t}$$

For $t > 0$

$$(0 + 4)v_p = 4(20) \implies v_p = 20\text{ V} \qquad t > 0$$

$$v = v_p + v_n = 20 + Ae^{-4t}$$

$$v(0^+) = 32 = 20 + Ae^0 \implies A = 12$$

$$v = 20 + 12e^{-4t}\text{ V} \qquad t > 0$$

20.

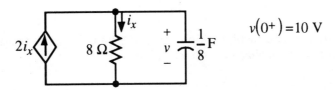

KCL gives

$$-2\left(\frac{1}{8}v\right) + \frac{1}{8}v + \frac{1}{8}\frac{d}{dt}v = 0$$

we rearrange this to get

$$\frac{d}{dt}v - v = 0$$

$$s - 1 = 0 \implies s = 1 = v = Ae^t$$

$$v(0^+) = 10 = Ae^0$$

$$v = 10e^t\text{ V} \qquad t > 0$$

No, the circuit is not stable

CHAPTER 9

Second-Order Circuits

Exercises

1.

Replace differentiation with multiplication by s and divide out the dependent variable to obtain the characteristic equation.

(a) $\dfrac{d^2}{dt^2}v + 6\dfrac{d}{dt}v + 8v = 0 \Rightarrow s^2 + 6s + 8 = 0$

$s = -2, -4 \Rightarrow \boxed{v_n = A_1 e^{-2t} + A_2 e^{-4t}}$

(b) $\dfrac{d^2}{dt^2}v + 4\dfrac{d}{dt}v + 4v = 0 \Rightarrow s^2 + 4s + 4 = 0$

$s = -2, -2 \Rightarrow \boxed{v_n = \left(A_1 + A_2 t\right)e^{-2t}}$

(c) $\dfrac{d^2}{dt^2}v + 6\dfrac{d}{dt}v + 25v = 0 \Rightarrow s^2 + 6s + 25 = 0$

$s = -3 + j4, s = -3 - j4 \Rightarrow \boxed{v = A_1 e^{(-3+j4)t} + A_2 e^{(-3-j4)t}}$

(d) $\dfrac{d^2}{dt^2}v + 4v = 0 \Rightarrow s^2 + 4s = 0$

$s = -j2, +j2 \Rightarrow \boxed{v = A_1 e^{j2t} + A_2 e^{-j2t}}$

2.

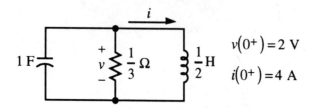

$v(0^+) = 2\ \text{V}$

$i(0^+) = 4\ \text{A}$

KCL gives

$$\frac{d}{dt}v + 3v + 2\int_{-\infty}^{t} v\,d\lambda = 0$$

Differentiation yields

$$\frac{d^2}{dt^2}v + 3\frac{d}{dt}v + 2v = 0 \;\Rightarrow\; s^2 + 3s + 2 = 0$$

$$s = -1, -2$$

$$v = A_1 e^{-t} + A_2 e^{-2t}$$

$$v(0^+) = 2 = A_1 e^0 + A_2 e^0 \;\Rightarrow\; A_1 + A_2 = 2$$

Differentiate v and substitute this into the original KCL equation. then, evaluate this at $t = 0^+$

$$-A_1 e^0 - 2A_2 e^0 + 3(2) + 4 = 0 \;\Rightarrow\; A_1 + 2A_2 = 10$$

Solve

$$\begin{bmatrix} 1 & 1 \\ 1 & 2 \end{bmatrix}\begin{bmatrix} A_1 \\ A_2 \end{bmatrix} = \begin{bmatrix} 2 \\ 10 \end{bmatrix} \;\Rightarrow\; A_1 = -6, \qquad A_2 = 8$$

$$\boxed{v = -6e^{-t} + 8e^{-2t}\text{ V} \qquad t > 0}$$

From the second equation

$$\begin{aligned} \omega_0^2 &= 2 \\ 2\zeta\omega_0 &= 3 \end{aligned} \;\Rightarrow\; \zeta = \frac{3}{2\omega_0} = \frac{3}{2\sqrt{2}} = \frac{3}{4}\sqrt{2}$$

3.

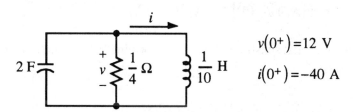

$$v(0^+) = 12\text{ V}$$
$$i(0^+) = -40\text{ A}$$

KCL gives

$$2\frac{d}{dt}v + 4v + 10\int_{-\infty}^{t} v\,d\lambda = 0$$

Differentiation and division by 2 yields

$$\frac{d^2}{dt^2}v + 2\frac{d}{dt}v + 5v = 0 \;\Rightarrow\; s^2 + 2s + 5 = 0$$

$$s = -1 + j2, \quad s = -1 - j2$$

$$v = e^{-t}(A\cos 2t + B\sin 2t)$$

$$v(0^+) = 12 = e^0(A\cos 0 + B\sin 0) \;\Rightarrow\; A = 12$$

Differentiate v and substitute this into the original KCL equation. Then evaluate this at $t = 0^+$.

$$2\left[e^0(-2A\sin 0 + 2B\cos 0) - e^0(A\cos 0 + B\sin 0)\right]$$

$$+4(12) - 40 = 0$$

$$4B - 2A + 8 = 0 \implies 4B - 24 + 8 = 0$$

$$B = 4$$

$$\boxed{v = e^{-t}(12\cos 2t + 4\sin 2t)\text{ V} \qquad t > 0}$$

4.

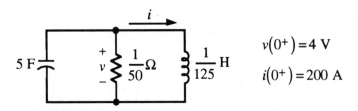

$$v(0^+) = 4\text{ V}$$
$$i(0^+) = 200\text{ A}$$

KCL gives

$$5\frac{d}{dt}v + 50v + 125\int_{-\infty}^{t} v\,d\lambda = 0$$

Differentiation and divide by 5

$$\frac{d^2}{dt^2}v + 10\frac{d}{dt}v + 25v = 0 \implies s^2 + 10s + 25 = 0$$

$$s = -5, \quad s = -5$$

$$v = (A_1 + A_2 t)e^{-5t}$$

$$v(0^+) = 4 = e^0\left[A_1 + A_2(0)\right] \implies A_1 = 4$$

Differentiate v and substitute this into the original KCL equation. Then evaluate this at $t = 0^+$.

$$5\left\{-5\left[A_1 + A_2(0)\right]e^0 + A_2 e^0\right\} + 50(4) + 200 = 0$$

$$-25(4) + 5A_2 + 200 + 200 = 0 \implies A_2 = -60$$

$$\boxed{v = (4 - 60t)e^{-5t}\text{ V} \qquad t > 0}$$

5.

$$i \quad 2\text{ H} \quad 10\ \Omega$$

$$\frac{1}{8}\text{ F} \quad v$$

$$i(0^+) = 12\text{ A}$$
$$v(0^+) = 36\text{ V}$$

KVL gives

$$2\frac{d}{dt}i+10i+8\int_{-\infty}^{t}id\lambda=0$$

Differentiation and divide by two:

$$\frac{d^2}{dt^2}i+5\frac{d}{dt}i+4i=0 \Rightarrow s^2+5s+4=0$$

$$s=-1, \quad s=-4$$

$$i=A_1e^{-t}+A_2e^{-4t}$$

$$i(0^+)=12=A_1e^0+A_2e^0 \Rightarrow A_1+A_2=12$$

Differentiate i and substitute this into the original KVL equation and evaluate this at $t=0^+$.

$$2\left[-A_1e^0-4A_2e^0\right]+10(12)+36=0$$

$$A_1+4A_2=78$$

$$\begin{bmatrix}1&1\\1&4\end{bmatrix}\begin{bmatrix}A_1\\A_2\end{bmatrix}=\begin{bmatrix}12\\78\end{bmatrix} \Rightarrow \begin{matrix}A_1=-10\\A_2=22\end{matrix}$$

$$\boxed{i=-10e^{-t}+22e^{-4t}\text{ A} \qquad t>0}$$

6.

$i(0^+)=4$ A

$v(0^+)=200$ V

KVL gives

$$5\frac{d}{dt}i+30i+125\int_{-\infty}^{t}id\lambda=0$$

Differentiation and division by 5 yields

$$\frac{d^2}{dt^2}i+6i+25i=0 \Rightarrow s^2+6s+25=0$$

$$s=-3+j4, \quad -3-j4$$

$$i=e^{-3t}(A\cos4t+B\sin4t)$$

$$i(0^+)=4=e^0(A\cos0+B\sin0) \Rightarrow A=4$$

Differentiate i and substitute this into the original KVL equation and evaluate this at $t=0^+$.

$$5\left[e^0(-4A\sin 0+4B\cos 0)-3e^0(A\cos 0+B\sin 0)\right]+30(4)+200=0$$

$$20B-15A+320=0 \quad \Rightarrow \quad 20B-60+320=0$$

$$B=-13$$

$$\boxed{i=e^{-3t}(4\cos 4t-13\sin 4t)\text{ A} \qquad t>0}$$

7.

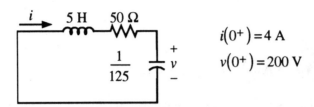

$$i(0^+)=4\text{ A}$$
$$v(0^+)=200\text{ V}$$

KVL gives

$$5\frac{d}{dt}i+50i+125\int_{-\infty}^{t}id\lambda=0$$

Differentiation and division by 5 yields

$$\frac{d^2}{dt^2}i+10\frac{d}{dt}i+25=0 \quad \Rightarrow \quad s^2+10s+25=0$$

$$s=-5,\ -5$$

$$i=\left(A_1+A_2 t\right)e^{-5t}\text{ A}, \qquad t>0$$

$$i(0^+)=4=\left[A_1+A_2(0)\right]e^0 \quad \Rightarrow \quad A_1=4$$

Differentiate i and substitute this into the original KVL equation and evaluate this at $t=0^+$.

$$5\left\{-5\left[A_1+A_2(0)\right]e^0+A_2 e^0\right\}+50(4)+200=0$$

$$-25(4)+5A_2+200+200=0 \quad \Rightarrow \quad A_2=-60$$

$$\boxed{i=(4-60t)e^{-5t}\text{ A} \qquad t>0}$$

8. (a)
$$\left(\frac{d^2}{dt^2}+6\frac{d}{dt}+8\right)v=12e^{-t}$$

Replace differentiation with multiplication by (-1).

$$\left[(-1)^2+6(-1)+8\right]v_p=12e^{-t}$$

$$\boxed{v_p=4e^{-t}\text{ V}}$$

(b)
$$\left(\frac{d^2}{dt^2}+6\frac{d}{dt}+25\right)v=68e^{-4t}$$

Replace differentiation with multiplication by (-4).

$$\left[(-4)^2+6(-4)+25\right]v_p=68e^{-t}$$

$$\boxed{v_p=4e^{-4t}\text{ V}}$$

(c) $\quad\left(\dfrac{d^2}{dt^2}+4\dfrac{d}{dt}+4\right)v_p=24$

$$\left[(0)^2+4(0)+4\right]v_p=24$$

$$\boxed{v_p=6\text{ V}}$$

(d) $\quad\left(\dfrac{d^2}{dt^2}+6\dfrac{d}{dt}+8\right)v=12e^{-2t}$

$$\left(\dfrac{d}{dt}+2\right)\left(\dfrac{d}{dt}+4\right)v=12e^{-2t}$$

$$v_p=\dfrac{1}{-2+4}t12e^{-2t}\ \Rightarrow\ \boxed{v_p=6te^{-2t}\text{ V}}$$

9.

$$v_s=\begin{cases}300\text{ V}&t<0\\150\text{ V}&t>0\end{cases}$$

The source is constant for $t<0$, so we can replace the inductance by a short circuit and the capacitance by an open circuit to calculate v_p and i_p for $t<0$.

$$v_p=300\text{ V}$$

$$i_p=-\dfrac{300}{50}=-6\text{ A}$$

$$v(0^+)=v(0^-)=v_p(0^-)=300\text{ V}$$

$$i(0^+)=i(0^-)=i_p(0^-)=-6\text{ A}$$

KCL for the top right-hand node of the original circuit is

$$\dfrac{1}{100}\dfrac{d}{dt}v+\dfrac{1}{50}v+\dfrac{1}{10}\int_{-\infty}^{t}(v-v_s)d\lambda=0$$

Differentiation and multiplication by 100 gives

$$\left(\frac{d^2}{dt^2}+2\frac{d}{dt}+10\right)v=10v_s$$

$$[0+2(0)+10]v_p=10(150) \implies v_p=150\text{ V} \qquad t>0$$

$$s^2+2s+10=0 \implies s=-1+j3,\ -1-j3$$

$$v=v_p+v_n=150+e^{-t}(A\cos 3t+B\sin 3t) \qquad t>0$$

$$v(0^+)=300=150+e^0(A\cos 0+B\sin 0) \implies A=150$$

Differentiate v, substitute it into the original KCL equation and evaluate this at $t=0^+$.

$$\frac{1}{100}\left[e^0(-3A\sin 0+3B\cos 0)-(A\cos 0+B\sin 0)\right]+\frac{1}{50}(300)-6=0$$

This reduces to

$$\frac{3}{100}B-\frac{1}{100}(150)+6-6=0 \implies B=50$$

$$\boxed{v=150+e^{-t}(150\cos 3t+50\sin 3t)\text{ V} \qquad t>0}$$

10.

$$v_s = 3e^{-4t}u(t)\text{ V}$$

First calculate the initial conditions. The source is zero for $t<0$, so

$$i(0^+)=i(0^-)=i_p(0^-)=0$$

$$v_c(0^+)=v_c(0^-)=v_{cp}(0^-)=0$$

For $t=0^+$, KVL gives

$$v(0^+)=-v_c(0^+)+v_s(0^+)=0+3=3\text{ V}$$

Application of KCL to the top right-hand node gives

$$\frac{d}{dt}(v-v_s)+3v+2\int_{-\infty}^{t}v\,d\lambda=0$$

Differentiation yields

$$\frac{d^2}{dt^2}v+3\frac{d}{dt}v+2v=\frac{d^2}{dt^2}3e^{-4t}$$

To find v_p, replace differentiation with multiplication by -4

$$\left[(-4)^2 + 3(-4) + 2\right]v_p = (-4)^2 3e^{-4t}$$

$$v_p = 8e^{-4t} \text{ V} \qquad t > 0$$

The characteristic equation is

$$s^2 + 3s + 2 = 0 \implies s = -1, \ -2$$

$$v = v_n + v_p = A_1 e^{-t} + A_2 e^{-2t} + 8e^{-4t} \qquad t > 0$$

$$v(0^+) = 3 = A_1 e^0 + A_2 e^0 + 8e^0 \implies A_1 + A_2 = -5$$

Differentiate v, substitute this into the original KCL equation and evaluate this at $t = 0^+$.

$$\left(-A_1 - 2A_2 - 4(8) - (-4)3\right) + 3(3) + 0 = 0$$

$$A_1 + 2A_2 = -11$$

$$\begin{bmatrix} 1 & 1 \\ 1 & 2 \end{bmatrix}\begin{bmatrix} A_1 \\ A_2 \end{bmatrix} = \begin{bmatrix} -5 \\ -11 \end{bmatrix} \implies \begin{bmatrix} A_1 \\ A_2 \end{bmatrix} = \begin{bmatrix} 1 \\ -6 \end{bmatrix}$$

$$\boxed{v = \left(e^{-t} - 6e^{-2t} + 8e^{-4t}\right)u(t) \text{ V}}$$

11.

$v_s = 0$ for $t < 0$, so

$$v(0^+) = v(0^-) = v_p(0^-) = 0$$

$$i(0^+) = i(0^-) = i_p(0^-) = 0$$

Replace inductance by a short circuit and capacitance by an open circuit to find

$$v_p = V_s \text{ V} \qquad t > 0$$

KVL gives

$$-v_s + \frac{d}{dt}i + 2i + 10\int_{-\infty}^t i\,d\lambda = 0$$

Differentiation yields

$$\left(\frac{d^2}{dt^2} + 2\frac{d}{dt} + 10\right)i = \frac{d}{dt}v_s$$

The characteristic equation is

$$s^2 + 2s + 10 = 0 \implies s = -1 + j3, \ -1 - j3$$

$$v = v_p + v_n = V_s + e^{-t}(A\cos 3t + B\sin 3t)$$

$$v(0^+) = 0 = V_s + e^0(A\cos 0 + B\sin 0) \implies A = -V_s$$

$$i(0^+) = \frac{1}{10}\frac{d}{dt}v\Big|_{t=0^+}$$

$$= e^0(-3A\sin 0 + 3B\cos 0) - (A\cos 0 + B\sin 0) = 0$$

$$B = \frac{1}{3}A = -\frac{1}{3}V_s$$

$$\boxed{v = \left[V_s - V_s e^{-t}\left(\cos 3t + \frac{1}{3}\sin 3t\right)\right]u(t)}$$

12.

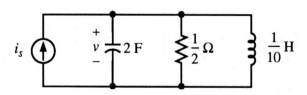

$$i_s = 30\cos 3t \text{ A}$$

KCL gives

$$-30\cos 3t + 2\frac{d}{dt}v + 2v + 10\int_{-\infty}^{t} v \, d\lambda = 0$$

Differentiation yields

$$\left(2\frac{d^2}{dt^2} + 2\frac{d}{dt} + 10\right)v = \frac{d}{dt}(30\cos 3t)$$

Replace $\cos 3t$ with e^{j3t}

$$\left(2\frac{d^2}{dt^2} + 2\frac{d}{dt} + 10\right)v_p = \frac{d}{dt}\left(30e^{j3t}\right)$$

To calculate v_p, replace differentiation with multiplication by $j3$

$$v_p = \frac{j3}{-8 + j6}30e^{j3t} = \frac{3e^{j90°}}{10e^{j143.13°}}\left(30e^{j3t}\right) = 9e^{j(3t-53.13°)}$$

$$v_p = \Re e\{v_p\} = 9\cos(3t - 53.13°)$$

13. $$\left(D^3 + 2D^2 + 3D + 1\right)v = 50e^{-2t}$$

$$\left[(-2)^3 + 2(-2)^2 + 3(-2) + 1\right] v_p = 50e^{-2t}$$

$$v_p = -10e^{-2t}$$

14. From Eq. (9.176)

$$v_2 = 1 + e^{-3t}\left(A_2 \cos t + B_2 \sin t\right) \qquad t > 0$$

Equation (9.166) gives

$$v_2(0^+) = 0$$

Therefore

$$v_2(0^+) = 0 = 1 + e^0\left(A_2 \cos 0 + B_2 \sin 0\right) \implies A_2 = -1$$

The KCL equation for node 2 is given by Eq. (9.168).

$$v_2 + 3\int_{-\infty}^{t}\left(v_2 - v_1\right)d\lambda = 0$$

This does not include a derivative of v_2, so differentiate the equation.

$$\frac{d}{dt}v_2 + 3\left(v_2 - v_1\right) = 0$$

Perform the differentiation and evaluate this at $t = 0^+$ $\left[v_1(0^+) = 0\right]$.

$$e^0\left(-A_2 \sin 0 + B_2 \cos 0\right) - 3\left(A_2 \cos 0 + B_2 \sin 0\right)$$

$$+3(0 - 0) = 0$$

$$B_2 = 3A_2 = -3$$

$$\boxed{v_2 = 1 - e^{-3t}\left(\cos t + 3\sin t\right) \text{ V} \qquad t > 0}$$

CHAPTER 10

The Sinusoidal Steady State

Exercises

1.   (a) $6\cos(200t + 50°)\,\text{mA} \Rightarrow 6\angle 50°\,\text{mA}$

       (b) $6\cos(400t + 50°)\,\text{mA} \Rightarrow 6\angle 50°\,\text{mA}$

       (c) $10\sin(100t + 30°)\,\text{mV} \Rightarrow 10\angle -60°\,\text{mV}$

2.   (a) $\mathbf{V} = \mathbf{ZI} = \left(0.5 \times 10^3 \angle 25°\right)\left(2.77 \times 10^{-3} \angle 50°\right)$

$$= 1.385 \angle 75°$$

$$\Rightarrow \boxed{V_m = 1.385\ \text{V},\ \phi_{\mathbf{V}} = 75°}$$

       (b) $\mathbf{V} = \mathbf{ZI} = \left((1 + j) \times 10^3\right)\left(100 \times 10^{-6} \angle 15°\right)$

$$= 141.4 \angle 30°\,\text{mV}$$

$$\Rightarrow \boxed{V_m = 141.4\ \text{mV},\ \phi_{\mathbf{V}} = 30°}$$

       (c) $\mathbf{V} = \mathbf{ZI} = 9e^{j30} \cdot (3 + j4)$

$$= 45 \angle 83.13°\,\text{V}$$

$$\Rightarrow \boxed{V_m = 45\ \text{V},\ \phi_{\mathbf{V}} = 83.13°}$$

3.   Define PHASOR: A phasor is a complex quantity used to represent the amplitude and phase of a sinsusoid.

        Define ROTATING PHASOR: A rotating phasor is a complex quantity representing a sinusoid of constant amplitude whose phase varies between $-\pi < \varphi < \pi$.

        Define IMPEDANCE: Impedance is a complex quantity representing the ratio of a voltage phasor to a current phasor.

4. (a) The equation $\mathbf{V} = \mathbf{ZI}$ was obtained by obtaining a differential equation relating $v(t)$ and $i(t)$. Then, the input was allowed to be a rotating phasor, $\mathbf{I}e^{j\omega t}$. It was shown that the particular response to this input was another rotating phasor, $\mathbf{V}e^{j\omega t}$. The resulting equation was then $\mathbf{V} = \mathbf{ZI}$.

    (b) The amplitude of $\mathbf{V}$, V_m, is then given by $V_m = |\mathbf{Z}|I_m$ and the phase of $\mathbf{V}$ is $\phi_{\mathbf{V}} = \phi_{\mathbf{Z}} + \phi_{\mathbf{I}}$.

5. (a)   (b)

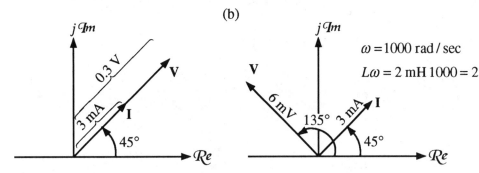

    (c)

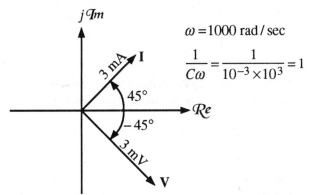

6. $v(t) = Ri(t) \Rightarrow A(j\omega) = 1$ and $B(j\omega) = R \Rightarrow \boxed{\mathbf{Z} = \dfrac{B(j\omega)}{A(j\omega)} = R}$

7. $v(t) = L\dfrac{d}{dt}i(t) \Rightarrow A(j\omega) = 1$ and $B(j\omega) = j\omega L \Rightarrow \boxed{\mathbf{Z} = \dfrac{B(j\omega)}{A(j\omega)} = j\omega L}$

8. $i(t) = C\dfrac{d}{dt}i(t)$: $i(t) = \mathbf{I}e^{j\omega t}$ and $v(t) = \mathbf{V}e^{j\omega t}$

    $\mathbf{I}e^{j\omega t} = C\dfrac{d}{dt}\mathbf{V}e^{j\omega t} = j\omega C\mathbf{V}e^{j\omega t} \Rightarrow \boxed{\mathbf{Z} = \dfrac{\mathbf{V}}{\mathbf{I}} = \dfrac{1}{j\omega C}}$

9.

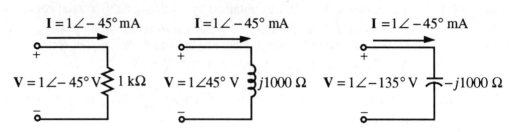

$$\mathbf{I} = 1\angle -45° \text{ mA} \qquad \mathbf{I} = 1\angle -45° \text{ mA} \qquad \mathbf{I} = 1\angle -45° \text{ mA}$$

$$\mathbf{V} = 1\angle -45° \text{ V} \gtrless 1 \text{ k}\Omega \quad \mathbf{V} = 1\angle 45° \text{ V} \gtrless j1000 \ \Omega \quad \mathbf{V} = 1\angle -135° \text{ V} \perp -j1000 \ \Omega$$

10. $\qquad v_R(t) = \cos(1000t - 45°) \text{ V} \ : \ v_L = \cos(1000t + 45°) \text{ V}$

$v_C(t) = \cos(1000t - 135°) \text{ V}$

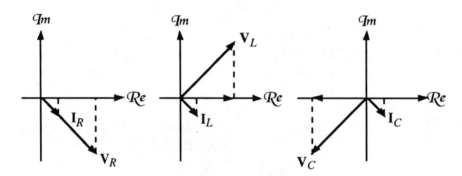

11. The practical significance of $\mathbf{V} = \mathbf{ZI}$ is that there is an algebraic method of relating the steady-state output voltage to an input current of sinusoidal form.

12. (a) $\mathbf{I} = 10\angle 20° \text{ mA} \quad \text{and} \quad \mathbf{V} = 50\angle 65° \text{ V}$

(b) $\mathbf{Z} = \dfrac{\mathbf{V}}{\mathbf{I}} = 5\angle 45° \text{ k}\Omega$

13. Given: $\dfrac{d}{dt}v + 2v = 4\dfrac{d}{dt}i + 4i$

If $i(t) = \mathbf{I}e^{j\omega t}$ and $v_p(t) = \mathbf{V}e^{j\omega t}$, then the above differential equation becomes

$$\frac{d}{dt}\mathbf{V}e^{j\omega t} + 2\mathbf{V}e^{j\omega t} = 4\frac{d}{dt}\mathbf{I}e^{j\omega t} + 4\mathbf{I}e^{j\omega t}$$

$$j\omega \mathbf{V}e^{j\omega t} + 2\mathbf{V}e^{j\omega t} = j\omega 4\mathbf{I}e^{j\omega t} + 4\mathbf{I}e^{j\omega t}$$

$$\mathbf{V}(2 + j3) = (4 + j12)\mathbf{I} \implies \boxed{\mathbf{Z} = \frac{\mathbf{V}}{\mathbf{I}} = 3.5\angle 15.26° \ \Omega}$$

14. $\mathbf{Z} = R + j\omega L + \dfrac{1}{j\omega C} = R + j\left(\omega L - \dfrac{1}{\omega C}\right)$

$$\implies |\mathbf{Z}| = \sqrt{R^2 + \left(\omega L - \frac{1}{\omega C}\right)^2}$$

$$\angle \mathbf{Z} = \text{arc } \tan\left(\frac{\omega L - \frac{1}{\omega C}}{R}\right)$$

Since $\mathbf{V} = \mathbf{ZI}$

$$\Rightarrow V_m = |\mathbf{Z}|I_m \quad \text{and} \quad \phi_V = \phi_I + \angle \mathbf{Z}$$

$$\text{then } v_p(t) = \sqrt{R^2 + \left(\omega L - \frac{1}{\omega C}\right)^2}\, I_m \cos\left\{\omega t + \phi_I + \text{arc } \tan\left[\frac{\omega L - (1/\omega C)}{R}\right]\right\}$$

15.

$$v(t) = Ri(t) + L\frac{di}{dt} + \frac{1}{C}\int_{-\infty}^{t} i(\lambda)d\lambda$$

$$\Rightarrow Dv(t) = \left(RD + LD^2 + \frac{1}{C}\right)i(t)$$

$$\Rightarrow j\omega\mathbf{V} = \left(Rj\omega + L(j\omega)^2 + \frac{1}{C}\right)\mathbf{I}$$

$$\mathbf{V} = \left(R + jL\omega + \frac{1}{jC\omega}\right)\mathbf{I}$$

On the other hand:

$$v_R(t) = Ri(t) \Rightarrow \mathbf{V}_R = R\mathbf{I}_m$$

$$v_L(t) = L\frac{di}{dt} \Rightarrow \mathbf{V}_L = Lj\omega\mathbf{I}_m$$

$$v_C(t) = \frac{1}{C}\int_{-\infty}^{t} i(\lambda)d\lambda \Rightarrow \mathbf{V}_C = \frac{1}{jC\omega}\mathbf{I}_m$$

Therefore:

$$\mathbf{V} = \mathbf{V}_R + \mathbf{V}_L + \mathbf{V}_C$$

16. (a) DRIVING-POINT IMPEDANCE: The ratio of a voltage phasor to a current phasor with the voltage phasor appearing across the same terminals as the current phasor.

(b) TRANSFER IMPEDANCE: The ratio of a voltage phasor to a current phasor with the voltage phasor not appearing across the same terminals as the current phasor.

(c) DRIVING-POINT ADMITTANCE: The ratio of a current phasor to a voltage phasor with the voltage phasor appearing across the same terminals as the current phasor.

(d)   TRANSFER ADMITTANCE: The ratio of a current phasor to a voltage phasor with the voltage phasor not appearing across the same terminals as the current phasor.

17.   (a)   RESISTANCE: The real part of an impedance, $\mathcal{R}e\{\mathbf{Z}\} = R(\omega)$.

       (b)   REACTANCE: The imaginary part of an impedance, $\mathcal{I}m\{\mathbf{Z}\} = X(\omega)$.

       (c)   CONDUCTANCE: The real part of an admittance, $\mathcal{R}e\{\mathbf{Y}\} = G(\omega)$.

       (d)   SUSCEPTANCE: The imaginary part of an admittance, $\mathcal{I}m\{\mathbf{Y}\} = B(\omega)$

18.   Given $\mathbf{Y} = 3 + j4 \Rightarrow \mathbf{Z} = \dfrac{1}{\mathbf{Y}} = \dfrac{1}{3 + j4} = 0.2\angle -53.13° = 0.12 - j0.16 \ \Omega$

       (a) $G = 3 \ \text{S}$

       (b) $B = 4 \ \text{S}$

       (c) $R = 0.12 \ \Omega$

       (d) $X = -0.16 \ \Omega$

19.   For series RLC circuit, the following differential equation applies.

$$v(t) = Ri(t) + L\frac{d}{dt}i(t) + \frac{1}{C}\int_{-\infty}^{t} i(\lambda)d\lambda \ \Rightarrow \ \frac{d}{dt}v(t) = L\frac{d^2}{dt^2}i(t) + R\frac{d}{dt}i(t) + \frac{1}{C}i(t)$$

If $i(t) = \mathbf{I}e^{j\omega t}$, then $v_p(t) = \mathbf{V}e^{j\omega t}$

$$j\omega\mathbf{V} = \left[(j\omega)^2 L + j\omega R + \frac{1}{C}\right]\mathbf{I} \ \Rightarrow \ \mathbf{V} = \left(R + j\omega L + \frac{1}{j\omega C}\right)$$

$$\Rightarrow \ \boxed{\mathbf{Z} = R + j\omega L + \frac{1}{j\omega C}} \ \Rightarrow \ \boxed{R(\omega) = R} \ : \ \boxed{X(\omega) = \omega L - \frac{1}{\omega C}}$$

$$\mathbf{Y} = \frac{1}{\mathbf{Z}} = \frac{1}{R + j\omega L + \dfrac{1}{j\omega C}} = \frac{j\omega C}{1 - \omega^2 LC + j\omega RC} \cdot \frac{(1 - \omega^2 LC - j\omega RC)}{(1 - \omega^2 LC - j\omega RC)}$$

$$\Rightarrow \ \boxed{\mathbf{Y} = \frac{j\omega C(1 - \omega^2 LC) + \omega^2 RC^2}{(1 - \omega^2 LC)^2 + (\omega RC)^2}} \ \Rightarrow \ \boxed{G(\omega) = \frac{\omega^2 RC^2}{(1 - \omega^2 LC)^2 + (\omega RC)^2}}$$

$$\boxed{B(\omega) = \frac{\omega C(1 - \omega^2 LC)}{(1 - \omega^2 LC)^2 + (\omega RC)^2}}$$

20.   (a)   Connect R_1 and R_2 in series. Then $\mathbf{Z} = R_1 + R_2$ and $\mathbf{Y} = \dfrac{1}{R_1 + R_2}$, which

illustrates that $\mathbf{Y} = \mathbf{Z}^{-1}$ for driving-point impedance and driving-point admittance.

(b)   Measure transfer impedance and admittance as shown below.

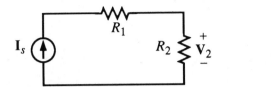

$$\mathbf{Z} = R_2$$

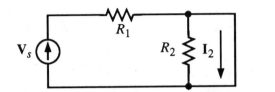

$$\mathbf{Y} = \frac{1}{R_1}$$

This illustrates that $\mathbf{Y} \neq \mathbf{Z}^{-1}$ for transfer admittance and impedance.

21.

(a)   KVL: $Ri(t) + L\dfrac{di}{dt} = v_s(t)$

$\Rightarrow (R + LD)i(t) = v_s(t)$

$\Rightarrow (R + LD)\underbrace{LDi(t)}_{v_o(t)} = LDv_s(t)$

$\Rightarrow (R + LD)v_o(t) = LDv_s(t)$

$\Rightarrow (D + R/L)v_o(t) = Dv_s(t)$

Let $\mathcal{A}(D) = D + R/L, \quad \mathcal{B}(D) = D$
then $\mathcal{A}(D)v_o(t) = \mathcal{B}v_s(t)$

(b) $\mathbf{H}(j\omega) = \dfrac{\mathbf{V}_o}{\mathbf{V}_s} = \left.\dfrac{\mathcal{B}(D)}{\mathcal{A}(D)}\right|_{D=j\omega} = \dfrac{j\omega}{R/L + j\omega}$

(c) $\mathbf{H}(j\omega) = \dfrac{j\omega}{R/L + j\omega} = \dfrac{j8}{20 + j8} = \dfrac{8\angle 90°}{21.54\angle 21.8°} = 0.3714\angle 68.2°$

$\Rightarrow v_o(t) = 3.714\cos(8t + 68.2°)\ \text{V}$

CHAPTER 11

AC Circuit Analysis

Exercises

1. $\mathbf{I} = 6 \angle 30°$

2. $\mathbf{V} = 5 \angle -150°$

3. $\mathbf{I} = 4 \angle -60°$

4. $\mathbf{V} = 3 \angle 120°$

5.   (a)   Voltage-controlled voltage source

$$v_2(t) = \mu_m v_1(t)$$

If $v_1(t) = \mathbf{V}_1 e^{j\omega t}$, then $v_2(t) = \mathbf{V}_2 e^{j\omega t} \Rightarrow \mathbf{V}_2 = \mu_m \mathbf{V}_1$

       (b)   Current-controlled voltage source

$$v_1(t) = r_m i_1(t)$$

If $i_1(t) = \mathbf{I}_1 e^{j\omega t}$, then $v_1 = \mathbf{V}_1 e^{j\omega t} \Rightarrow \mathbf{V}_2 = r_m \mathbf{I}_1$

       (c)   Voltage-controlled current source

$$i_2(t) = g_m v_1(t)$$

If $v_1(t) = \mathbf{V}_1 e^{j\omega t}$, then $i_2(t) = \mathbf{I}_2 e^{j\omega t} \Rightarrow \mathbf{I}_2 = g_m \mathbf{V}_1$

100

(d)   Currrent-controlled current source

$$i_2(t) = \alpha_m i_1(t)$$

If $i_1(t) = I_1 e^{j\omega t}$, then $i_2(t) = I_2 e^{j\omega t} \Rightarrow I_2 = \alpha_m I_1$

6.   Using the law of cosines

$$I_{m1} = \sqrt{I_{m2}^2 + I_{m3}^2 - 2I_{m2}I_{m3}\cos(\angle I_3 - \angle I_2)}$$

7.   Figure (a) is the time-domain form of the circuit, involving real functions of time. In general, solutions in the time domain require the use of differential equations. Figure (e) is the frequency-domain form of the circuit. Functions of time are assumed to be complex exponentials of a given frequency. Solutions rely on algebraic techniques already developed for resistive circuit analysis.

8.
$$H(j\omega) = \frac{3j\omega}{1 - \omega^2 + j\omega} = \frac{V_o}{V_s}$$

$$(1 - \omega^2 + j\omega)V_o = 3j\omega V_s$$

$$(D^2 + D + 1)v_o(t) = 3Dv_s(t)$$

$$\Rightarrow \frac{d^2}{dt^2}v_o(t) + \frac{d}{dt}v_o(t) + v_o(t) = 3\frac{d}{dt}v_s(t)$$

9.
$$Z_1 \parallel Z_2 = \frac{1}{\frac{1}{Z_1} + \frac{1}{Z_2}} = \frac{Z_1 Z_2}{Z_1 + Z_2} = \frac{Z_2 Z_1}{Z_1 + Z_2} = Z_2 \parallel Z_1$$

$$(Z_1 \parallel Z_2) \parallel Z_3 = \frac{1}{\frac{1}{Z_1} + \frac{1}{Z_2} + \frac{1}{Z_3}} = Z_1 \parallel (Z_2 \parallel Z_3)$$

10.   In order to prove the current divider equation, refer to the circuit shown below.

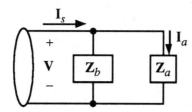

KCL: $\dfrac{V}{Z_a} + \dfrac{V}{Z_b} = I_s$, where $V = Z_a I_a$

Then, $\left(1 + \dfrac{Z_a}{Z_b}\right)I_a = I_s \Rightarrow \boxed{I_a = \dfrac{Z_b}{Z_a + Z_b}I_s}$

In order to prove the voltage divider equation, refer to the circuit shown below.

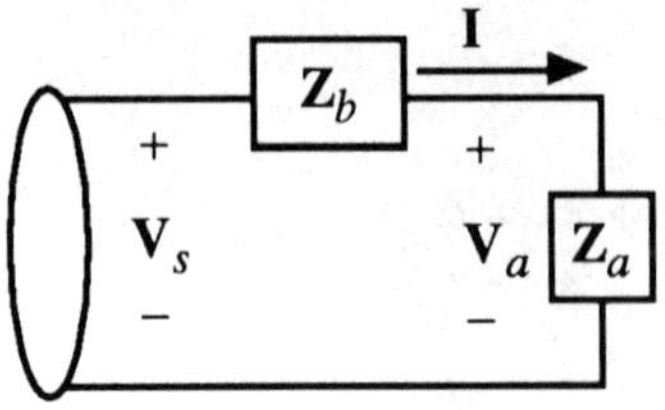

KVL: $\left(\mathbf{Z}_a + \mathbf{Z}_b\right)\mathbf{I} = \mathbf{V}_s$, where $\mathbf{I} = \dfrac{\mathbf{V}_a}{\mathbf{Z}_a} \Rightarrow \mathbf{V}_a\left(1 + \dfrac{\mathbf{Z}_b}{\mathbf{Z}_b}\right) = \mathbf{V}_s$

$$\Rightarrow \boxed{\mathbf{V}_a = \frac{\mathbf{Z}_a}{\mathbf{Z}_a + \mathbf{Z}_b}\mathbf{V}_s}$$

11.　　　　Given the circuit below:

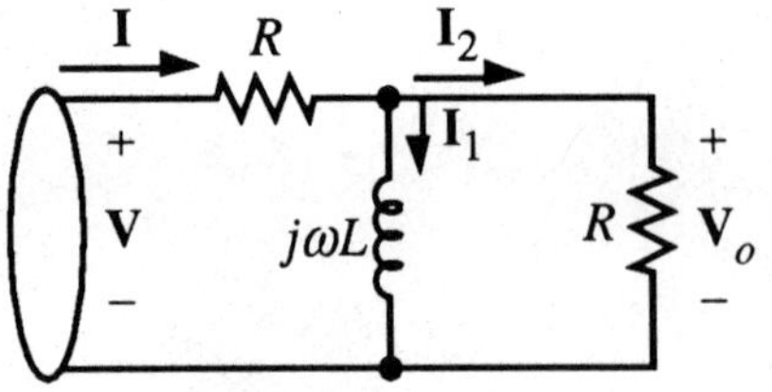

(a)　　$\boxed{\mathbf{Z} = R + \dfrac{j\omega RL}{R + j\omega L}}$

(b)　　Voltage divider: $\mathbf{V}_o = \dfrac{j\omega L \,\|\, R}{R + \left(j\omega L \,\|\, R\right)}\mathbf{V} \Rightarrow \boxed{\dfrac{\mathbf{V}_o}{\mathbf{V}} = \dfrac{j\omega L}{R + j\omega 2L}}$

(c)　　$\mathbf{I}_1 = \dfrac{\mathbf{V}_o}{j\omega L} \Rightarrow \dfrac{\mathbf{I}_1}{\mathbf{V}} = \dfrac{1}{j\omega L}\dfrac{\mathbf{V}_o}{\mathbf{V}} \Rightarrow \boxed{\dfrac{\mathbf{I}_1}{\mathbf{V}} = \dfrac{1}{R + j\omega 2L}}$

(d)　　$\mathbf{I}_2 = \dfrac{\mathbf{V}_o}{R} \Rightarrow \dfrac{\mathbf{I}_2}{\mathbf{V}} = \dfrac{1}{R}\dfrac{\mathbf{V}_o}{\mathbf{V}} \Rightarrow \boxed{\dfrac{\mathbf{I}_2}{\mathbf{V}} = \dfrac{j\omega L}{R^2 + j\omega 2RL}}$

(e)　　$\dfrac{\mathbf{I}_1}{\mathbf{I}} = \dfrac{R}{R + j\omega L}$

(f)　　$\dfrac{\mathbf{I}_2}{\mathbf{I}} = \dfrac{j\omega L}{R + j\omega L}$

12.　　　　Given $\dfrac{\mathbf{V}}{\mathbf{I}} = \dfrac{-\omega^2 + j\left(\omega / R_1 C\right)}{-\omega^2 + \left(\dfrac{R_1 + R_2}{L}\right)\left(\dfrac{1}{R_1 C}\right) + j\omega\left(\dfrac{1}{R_1 C} + \dfrac{R_2}{L}\right)} R_2,$

as $\omega \to \infty, \dfrac{V}{I} \to R_2$. This is correct since the capacitor will become a short circuit as $\omega \to \infty$ and the inductor will become an open circuit. Then $V = R_2 I \Rightarrow Z = R_2$.

13.

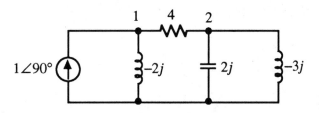

KCL node 1: $-1\angle 90° - 2jV_1 + 4(V_1 - V_2) = 0$

KCL node 2; $4(V_2 - V_1) + 2jV_2 - 3jV_2 = 0$

$$\begin{bmatrix} 4-2j & -4 \\ -4 & 4-j \end{bmatrix}\begin{bmatrix} V_1 \\ V_2 \end{bmatrix} = \begin{bmatrix} 1\angle 90° \\ 0 \end{bmatrix}$$

$$\Rightarrow \begin{array}{l} V_1 = 0.34\angle 175.43° \text{ V} \\ V_2 = 0.33\angle -170.54° \text{ V} \end{array}$$

14. Given the circuit shown below:

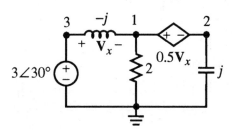

Node 1-2, KVL: $V_1 - V_2 = 0.5(3\angle 30° - V_1)$

Node 1-2, KCL: $2V_1 + jV_2 - j(V_1 - 3\angle 30°) = 0$

$$\begin{bmatrix} 1.5 & -1 \\ 2-j & j \end{bmatrix}\cdot\begin{bmatrix} V_1 \\ V_2 \end{bmatrix} = \begin{bmatrix} 1.5\angle 30° \\ 3\angle -60° \end{bmatrix} \Rightarrow \begin{bmatrix} V_1 \\ V_2 \end{bmatrix} = \begin{bmatrix} 0.727\angle -74.04° \\ 2.059\angle -119.04° \end{bmatrix}$$

15.

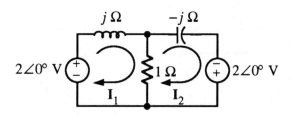

KVL mesh 1: $-2\angle 0° + jI_1 + (I_1 - I_2) = 0$

KVL mesh 2: $I_2 - I_1 + -jI_2 - 2\angle 0° = 0$

$$\begin{bmatrix} 1+j & -1 \\ -1 & 1-j \end{bmatrix}\begin{bmatrix} \mathbf{I}_1 \\ \mathbf{I}_2 \end{bmatrix} = \begin{bmatrix} 2\angle 0° \\ 2\angle 0° \end{bmatrix}$$

$$\Rightarrow\ \mathbf{I}_1 = 4.47\angle -26.57°\ \text{A}$$
$$\mathbf{I}_2 = 4.47\angle 26.57°\ \text{A}$$

16.   Given the circuit shown below:

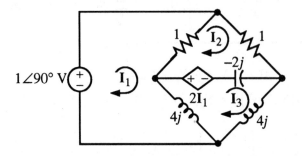

Mesh 1 KVL: $(1+j4)\mathbf{I}_1 - \mathbf{I}_2 - j4\mathbf{I}_3 = j$

Mesh 2 KVL: $-\mathbf{I}_1 + (2-j2)\mathbf{I}_2 - 2\mathbf{I}_1 + j2\mathbf{I}_3 = 0$

Mesh 3 KVL $(2-j4)\mathbf{I}_1 + j2\mathbf{I}_2 + j6\mathbf{I}_3 = 0$

$$\begin{bmatrix} 1+4j & -1 & -4j \\ 1 & 2-2j & 2j \\ 2-4j & 2j & 6j \end{bmatrix}\begin{bmatrix} \mathbf{I}_1 \\ \mathbf{I}_2 \\ \mathbf{I}_3 \end{bmatrix} = \begin{bmatrix} j \\ 0 \\ 0 \end{bmatrix}$$

$$\Rightarrow\ \mathbf{I}_1 = 0.49\angle 14.04°\ \text{A}$$
$$\mathbf{I}_2 = 0.57\angle 47.18°\ \text{A}$$
$$\mathbf{I}_3 = 0.17\angle 33.48°\ \text{A}$$

17.   Given the circuit shown below:

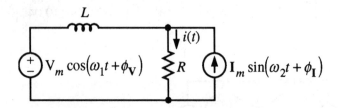

Keep the ω_1 source:

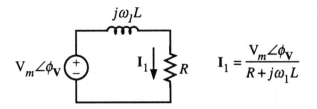

$$I_1 = \frac{V_m \angle \phi_V}{R + j\omega_1 L}$$

$$i_1(t) = \frac{V_m}{\sqrt{R^2 + (\omega_1 L)^2}} \cos(\omega_1 t + \phi_v - \theta_1)\,A \quad \text{where } \theta_1 = \text{arc } \tan\left(\frac{\omega_1 L}{R}\right)$$

Keep the ω_2 source:

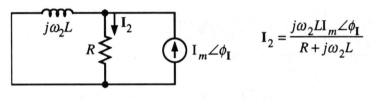

$$I_2 = \frac{j\omega_2 L I_m \angle \phi_I}{R + j\omega_2 L}$$

$$i_2(t) = \frac{\omega_2 L I_m}{\sqrt{R^2 + (\omega_2 L)^2}} \cos(\omega_2 t + \phi_I + 90° - \theta_2)\,A \quad \text{where } \theta_2 = \text{arc } \tan\left(\frac{\omega_2 L}{R}\right)$$

$$i(t) = \frac{V_m}{\sqrt{R^2 + (\omega_1 L)^2}} \cos(\omega_1 t + \phi_v - \theta_1) + \frac{\omega_2 L I_m}{\sqrt{R^2 + (\omega_2 L)^2}} \cos(\omega_2 t + \phi_I + 90 - \theta_2)\,A$$

18.　　　　Given the circuit shown below:

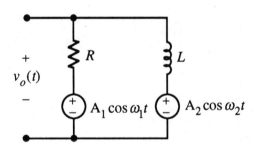

Keep the ω_1 source:

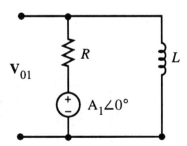

$$\mathbf{V}_{01} = \frac{jL\omega_1}{R + jL\omega_1} \cdot A_1 \angle 0° \implies v_{01}(t) = B_1 \cos(\omega_1 t + \phi_1)$$

where $B_1 = \dfrac{\omega_1 L A_1}{\sqrt{R^2 + (L\omega_1)^2}}$, $\phi_1 = 90° - \arctan\left(\dfrac{L\omega_1}{R}\right)$

Keep the ω_2 source:

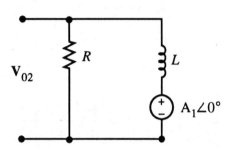

$$\mathbf{V}_{02} = \frac{R}{R + jL\omega_2} \cdot A_2 \angle 0° \implies v_{02}(t) = B_2 \cos(\omega_2 t + \phi_2)$$

where $B_2 = \dfrac{R A_2}{\sqrt{R^2 + (L\omega_2)^2}}$, $\phi_2 = -\arctan\left(\dfrac{L\omega_2}{R}\right)$

$$v_0(t) = v_{01}(t) + v_{02}(t) = B_1 \cos(\omega_1 t + \phi_1) + B_2 \cos(\omega_2 t + \phi_2).$$

19.   Given the circuit shown below:

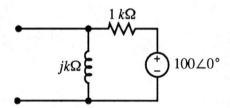

$$\mathbf{V}_{oc} = \frac{j}{1+j} 100 \angle 0° = \frac{100}{\sqrt{2}} \angle 45° \text{ V}$$

$$\mathbf{Z}_{TH} = j \| 1 = \frac{j}{1+j} = \frac{1}{\sqrt{2}} \angle 45° \text{ k}\Omega$$

$$\mathbf{I}_{SC} = \frac{\mathbf{V}_{oc}}{\mathbf{Z}_{TH}} = 100 \angle 0° \text{ mA}$$

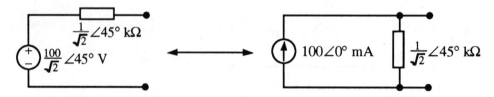

20.   Given the circuit shown below:

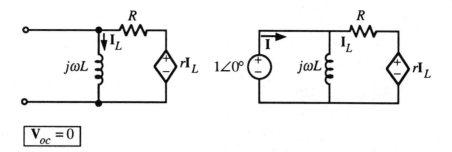

$$\boxed{V_{oc} = 0}$$

Apply a $1\angle0°$ voltage soucre as shown and solve for **I**.

KCL: $\mathbf{I} = \dfrac{1}{j\omega L} + \dfrac{1 - r(1/j\omega L)}{R} = \dfrac{R - r + j\omega L}{j\omega RL}$ $\Rightarrow$ $\boxed{\mathbf{Z}_{TH} = \dfrac{j\omega RL}{R - r + j\omega L} = \mathbf{Z}_N}$

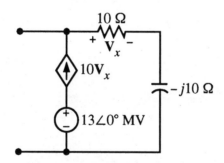

THÉVENIN and NORTON

21.   Given the circuit shown below:

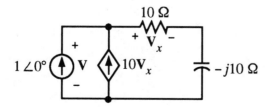

$\mathbf{V}_x = (10\mathbf{V}_x)10 = 100\mathbf{V}_x \Rightarrow \mathbf{V}_x = 0$

$\Rightarrow \mathbf{V}_{oc} = \mathbf{V}_x - j10(10\mathbf{V}_x) = 0$

Find $\mathbf{Z}_{TH}$:

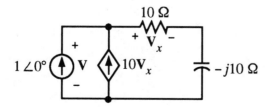

$$\mathbf{V}_x = 10\left(1\angle 0° + 10\mathbf{V}_x\right)$$

$$\Rightarrow \mathbf{V}_x = -\frac{10}{99}$$

$$\mathbf{V} = \mathbf{V}_x - j10\left(1 + 10\mathbf{V}_x\right)$$

$$= \left(1 - j100\right)\left(-\frac{10}{99}\right) - j10$$

$$= -\frac{10}{99} + j\frac{10}{99} = \frac{10\sqrt{2}}{99}\angle 135° \text{ V}$$

$$\Rightarrow \mathbf{Z}_{TH} = \frac{10\sqrt{2}}{99}\angle 135° \text{ }\Omega$$

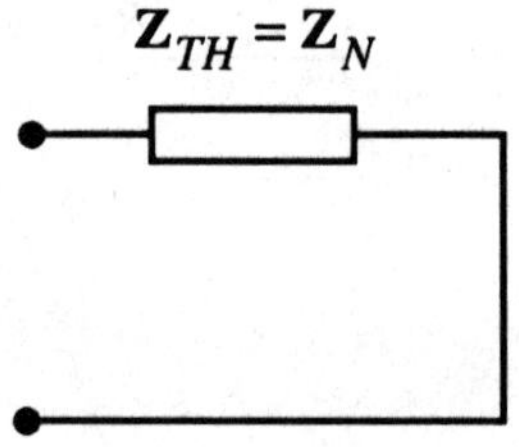

$$\mathbf{Z}_{TH} = \mathbf{Z}_N$$

CHAPTER 12

Power In AC Circuits

Exercises

1. (a) $p(t) = v(t)i(t)$ W

    (b) $P = \dfrac{1}{2}V_m I_m \cos(\phi_V - \phi_I) = \dfrac{1}{2}V_m I_m \cos(\theta)$ W

    (c) $PF = \cos(\phi_V - \phi_I) = \cos(\theta)$ (unitless)

2. $I_m = 10$ A, $\mathbf{Z} = 10\angle -45°\ \Omega$

    (a) $PF = \cos(-45°) = \boxed{0.707\ \text{leading}}$

    (b) $P = \dfrac{1}{2}V_m I_m \cos(\theta) = \dfrac{1}{2}(10)(10)(10)(0.7071) = \boxed{353.6\ \text{W}}$

3. P is nonnegative for a passive load. Therefore, $\cos(\theta) \geq 0$.

    $$\Rightarrow\ -\frac{\pi}{2} \leq \theta \leq \frac{\pi}{2}\ \Rightarrow\ R(\omega) = \mathcal{R}e\{\mathbf{Z}(j\omega)\} \geq 0$$

4. $P_{app} = \dfrac{P}{PF} = \dfrac{1}{2}V_m I_m$ VA

5. $I_m = 10$ A, $\mathbf{Z} = 10\angle -45°\ \Omega$

    $$P_{app} = \frac{1}{2}V_m I_m = \frac{1}{2}(10)(10)(10) = \boxed{500\ \text{VA}}$$

6.   See the section on apparent power for practical significance.

7. $Q = \dfrac{1}{2}V_m I_m \sin(\theta)$ VAR

8. Phase quadrature implies a 90° phase shift between two signals. Q is appropriate to remind us that the reactive power is due to the part of the voltage $v(t)$ that is in phase quadrature with the current $i(t)$.

9. $$I_m = 10 \text{ A}, \quad \mathbf{Z} = 10\angle -45° \; \Omega$$

    $$Q = \frac{1}{2} V_m I_m \sin(-45°) = \frac{1}{2}(10)(10)(10)(-0.7071) = \boxed{-353.6 \text{ VAR}}$$

10. $$\mathbf{S} = \frac{1}{2} \mathbf{VI}^* \text{ VA}$$

11.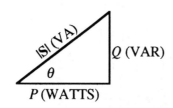

12. $$I_m = 10 \text{ A}, \quad \mathbf{Z} = 10\angle -45° \; \Omega$$

    (a) $$\mathbf{S} = \frac{1}{2}\mathbf{VI}^* = \frac{1}{2}(100\angle -45°)(10\angle 0°) = 500\angle -45°. \quad \text{VA} = \boxed{353.6 - j353.6 \text{ VA}}$$

    (b)
    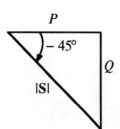

    (c) $$P = |\mathbf{S}|\cos(-45°) = 500 \times 0.7071 = \boxed{353.6 \text{ W}}$$

    $$Q = |\mathbf{S}|\sin(-45°) = 500 \times (-0.7071) = \boxed{-353.6 \text{ VAR}}$$

    $$P_{app} = |\mathbf{S}| = \boxed{500 \text{ VA}}$$

13. $$I_{rms} = \frac{V_{rms}}{|\mathbf{Z}|} = \frac{120}{390} = 0.308 \text{ A}_{rms}$$

    $$P = I_{rms}^2 R(\omega) = (0.308)^2 (100) = 9.47 \text{ W}$$

    $$Q = I_{rms}^2 X(\omega) = (0.308)^2 (377) = 35.7 \text{ VAR}$$

    $$\theta = 75.1°$$

14. (a) Take the real parts of both side of Eq. (12.47c):

$$\sum_{\substack{\text{All} \\ \text{components}}} P = 0$$

The average power supplied by the sources in an ac network equals the average power delivered to the passive components in the network.

(b) As in part (a)

$$\sum_{\substack{\text{All} \\ \text{components}}} Q = 0$$

The reactive power supplied by the sources in an ac network equals the reactive power delivered to the passive components in the network.

15. (a) $P = \dfrac{5 \times 745.5}{0.7} = \boxed{5.33 \text{ kW}}$

$\cos\theta = 0.8, \quad \theta > 0 \implies \tan\theta = 0.75$

$Q = P \tan\theta = \boxed{4 \text{ kVAR}}$

(b) $(\cos\theta)_{\text{new}} = 0.95 \implies (\tan\theta)_{\text{new}} = 0.33$

$Q_{\text{new}} = P(\tan\theta)_{\text{new}} = 1.75 \text{ kVAR}$

$Q_{\text{new}} - Q = 1.75 - 4 = \boxed{-2.25 \text{ kVAR}}$

(c) $\dfrac{1}{2} V_m^2 (C\omega) = 2250$

$\implies C = \dfrac{2250}{480^2 \times 2\pi \times 60} = \boxed{25.9\,\mu\text{F}}$

16. $\mathbf{Z}_s = \dfrac{\mathbf{V}_{oc}}{\mathbf{I}_{sc}} = \dfrac{10\angle -65^\circ}{2\angle 20^\circ} = 5\angle -85^\circ \ \Omega = 0.436 + j4.98 \ \Omega \implies R_s = 3.54 \ \Omega$

$P_a = \max\{P_L\} = \dfrac{V_{sm}^2}{8R_s} = \dfrac{(10)^2}{8(0.436)} \implies \boxed{P_a = 28.67 \text{ W}}$

17. $\mathbf{Z}_s = 1 \| j = \dfrac{1}{2} + j\dfrac{1}{2} \;\Rightarrow\; R_s = \dfrac{1}{2}\,\Omega$

$$\boxed{\;\max\{P_L\} = \frac{V_{sm}^2}{8R_s} = \frac{(100)^2}{8(0.5)} = 2500\ \text{W}\;}$$

18. $\mathbf{Z} = \dfrac{1}{2} - j\dfrac{1}{2}\,\Omega$

19. $\max\{P_L\} = \dfrac{1}{2}\dfrac{V_{sm}^2}{R_s} = \dfrac{1}{2}\dfrac{(100)^2}{(0.5)} = 10{,}000\ W$

20. $\mathbf{Z} = 0 - j\dfrac{1}{2}\,\Omega$

CHAPTER 13

Frequency Response

Exercises

1. $$\mathbf{H}(j\omega) = \frac{j\omega L}{R + j\omega L}$$

If $\omega = \frac{R}{L}$, $\mathbf{H}\left(j\frac{R}{L}\right) = \frac{j(R/L)L}{R + j(R/L)L} = \frac{jR}{R + jR}$ $\Rightarrow$ $\boxed{\mathbf{H}\left(j\frac{R}{L}\right) = \frac{1}{\sqrt{2}}\angle 45°}$

2. $$\mathbf{V}_{o1} = \frac{j(10^3\pi)(7\times 10^{-3})}{2200 + j(10^3\pi)(7\times 10^{-3})}(3\angle 10°) \Rightarrow \mathbf{V}_{o1} = 0.0299\angle 99.4°$$

$$\Rightarrow v_{o1} = 0.0299\cos(10^3\pi t + 99.4°)\,\text{V}$$

$$\mathbf{V}_{o2} = \frac{j(10^7\pi)(7\times 10^{-3})}{2200 + j(10^7\pi)(7\times 10^{-3})}(3\angle 27°) \Rightarrow \mathbf{V}_{o2} = 0.299\angle 27.57°$$

$$\Rightarrow v_{o2} = 0.299\cos(10^7\pi t + 27.57°)\,\text{V}$$

$$\Rightarrow \boxed{v_o(t) = 0.0299\cos(10^3\pi t + 99.4°) + 0.299\cos(10^7\pi t + 27.5°)\,\text{V}}$$

3. Case 1: $P_{av} = 1\,\text{W} = \dfrac{1}{2}\dfrac{V_m^2}{R} = \dfrac{1}{2}\dfrac{A_1^2}{R}$ $\Rightarrow R = \dfrac{A_1^2}{2}$

    Case 2: $P_{av} = \dfrac{1}{2}\dfrac{A_2^2}{R} = \dfrac{1}{2}\cdot\dfrac{A_1^2/2}{A_1^2/2} = \boxed{\dfrac{1}{2}\,\text{W}}$

4.   Given the circuit shown below:

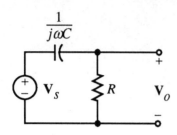

$$H(j\omega) = \frac{j\omega RC}{1 + j\omega RC}$$

(a) $|H(j\omega)| = \dfrac{\omega RC}{\sqrt{1+(\omega RC)^2}}$: $\angle H(j\omega) = 90° - \tan^{-1}(\omega RC)$

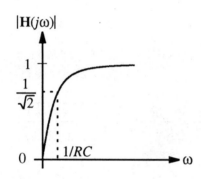

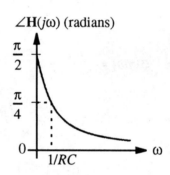

(b) $\omega_o = \dfrac{1}{RC}$

(c)   Stopband: $\omega < \omega_o$
        Passband: $\omega > \omega_o$

5.   Given the circuit shown below

(a) $H(j\omega) = \dfrac{1}{\sqrt{1+(\omega L / R)^2}} \angle -\tan^{-1}\left(\dfrac{\omega L}{R}\right)$

(b) $\omega_o = \dfrac{R}{L}$

(c)   Stopband: $\omega > \omega_o$
        Passband: $0 \le \omega \le \omega_o$

6.

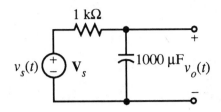

For $v_{s1}(t) = 10,\ v_{o1}(t) = v_{s1}(t) = 10\ \text{V}$

For $v_{s2}(t) = 10\cos 0.1t$

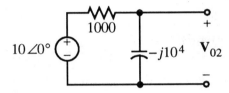

$$\mathbf{V}_{o2} = \frac{-j10^4}{10^3 - j10^4} \cdot 10\angle 0° = 9.96\angle -5.71°$$

$$\Rightarrow v_{o2}(t) = 9.96\cos(0.1t - 5.71)\ \text{V}$$

For $v_{s3}(t) = 10\cos t$

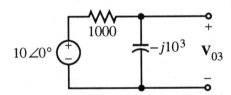

$$\mathbf{V}_{o3} = \frac{-j10^3}{10^3 - j10^3} \cdot 10\angle 0° = 5\sqrt{2}\angle -45°$$

$$\Rightarrow v_{o3}(t) = 5\sqrt{2}\cos(t - 45°)\ \text{V}$$

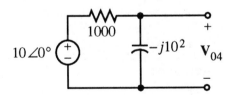

For $v_{s4}(t) = 10\cos 10t$

$$\mathbf{V}_{o4} = \frac{-j10^2}{10^3 - j10^2} \cdot 10\angle 0° = 0.995\angle -84.29°$$

$$\Rightarrow v_{o4}(t) = 0.995\cos(10t - 84.29°)\ \text{V}$$

$$v_{o(t)} = v_{o1}(t) + v_{o2}(t) + v_{o3}(t) + v_{o4}(t)$$

$$= \left[10 + 9.96\cos(0.1t - 5.71°) + 5\sqrt{2}\cos(t - 45°) + 0.995\cos(10t - 84.29°)\right]\ \text{V}$$

7.

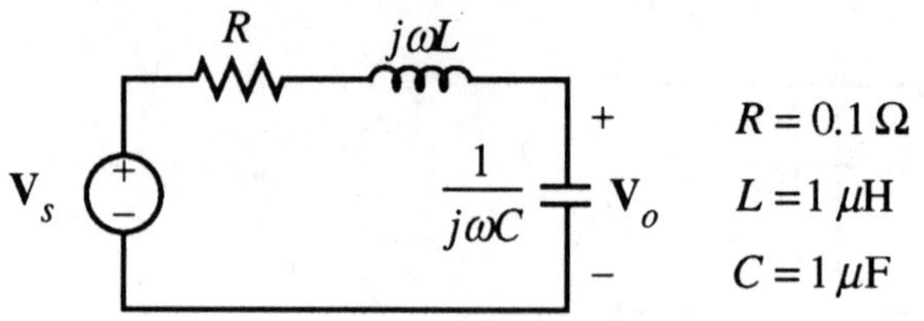

For $v_{s1}(t) = 10$, $v_{o1}(t) = 10$

For $v_{s2}(t) = 7\cos 10^6 t$, $\omega = 10^6$, $\mathbf{V}_{s2} = 7\angle 0°$

By Fig. 13.7,
$$H(j\omega) \cong 10 \angle -90°$$
$$\Rightarrow v_{o2}(t) \cong 70\cos(10^6 t - 90°)$$

For $v_{s3}(t) = 3\sin(5\times 10^6 t)$, $\omega = 5\times 10^6$, $\mathbf{V}_{s3} = 3\angle -90°$

By Fig. 13.7,
$$H(j\omega) \cong 0$$
$$\Rightarrow v_{o3}(t) \cong 0$$

$$v_o(t) = v_{o1}(t) + v_{o2}(t) + v_{o3}(t)$$
$$\cong 10 + 70\cos(10^6 t - 90°) \text{ V}$$

8.

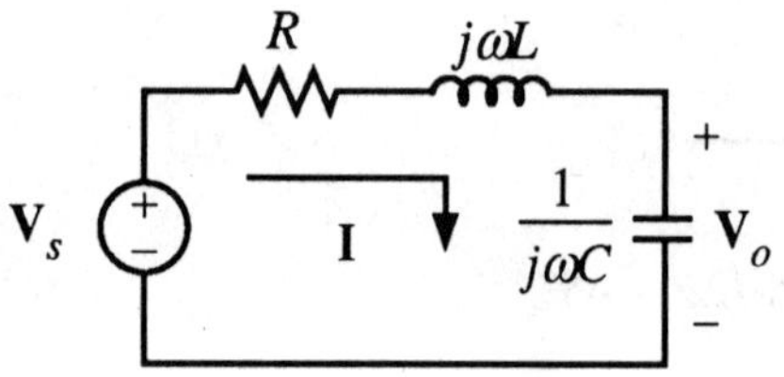

(a) $\quad \mathbf{Z}(j\omega) = R + j\omega L + \dfrac{1}{j\omega C}$

(b) $\quad |\mathbf{I}| = \dfrac{|\mathbf{V}_s|}{|\mathbf{Z}(j\omega)|}$

$\Rightarrow |\mathbf{I}|$ is maximum when $|\mathbf{Z}(j\omega)|$ is minimum, that is when $j\omega L + \dfrac{1}{j\omega C} = 0$.

(c) $\quad |\mathbf{I}|_{max}$ occurs when $j\omega L + \dfrac{1}{j\omega C} = 0 \Rightarrow \omega_{mr} = \dfrac{1}{\sqrt{LC}}$.

(d) $\quad \omega_{mr} = \dfrac{1}{\sqrt{LC}} = \dfrac{1}{\sqrt{10^{-6}\times 10^{-6}}} = 10^6 \text{ rad/sec}$

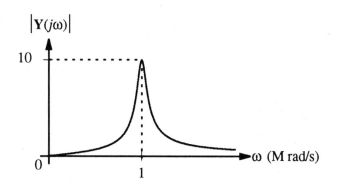

The plot of $\mathbf{Y}(j\omega)$ looks similiar to the plot of Fig. 13.7a for $\left|\mathbf{H}(j\omega)\right|$.

9.   Given the circuit shown below:

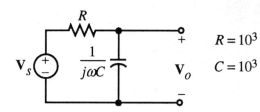

$R = 10^3$
$C = 10^3$

$A(s) = 1 + 2 = 0 \Rightarrow s = -1$

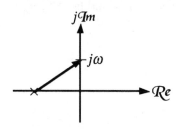

$$\mathbf{H}(j\omega) = \frac{1}{1 + j\omega}$$

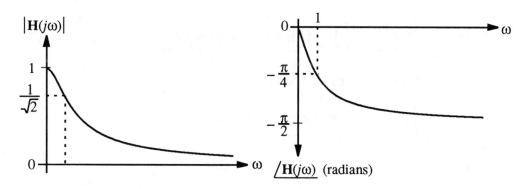

10.   Given the circuit shown below:

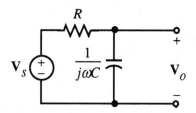

$$H(j\omega) = \frac{\frac{1}{RC}}{j\omega + \frac{1}{RC}}$$

(a)

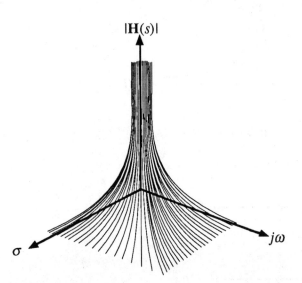

|H(s)|

σ   jω

(b) $\left|\mathbf{H}(j\omega)\right|$ is the trace of $\left|\mathbf{H}(s)\right|$ in the $j\omega$ plane with $\sigma = 0$. Imagine cutting through the 3-D plot along the $\sigma = 0$ axis. The result would be $\left|\mathbf{H}(j\omega)\right|$.

11.   Given the circuit shown below:

$i(t)$

+

$v(t)$ $1\,\Omega$ $1\,\text{mH}$

−

(a) $i(t) = \frac{v(t)}{R} + \frac{1}{L}\int_{-\infty}^{t} v(\lambda)d\lambda \;\Rightarrow\; \frac{1}{R}\frac{d}{dt}v + \frac{v}{L} = \frac{d}{dt}i$

If $i(t) = \mathbf{I}e^{st}$, then $v(t) = \mathbf{V}e^{st}$.

$\frac{s}{R}\mathbf{V} + \frac{\mathbf{V}}{L} = s\mathbf{I} \;\Rightarrow\; \boxed{\mathbf{Z} = \frac{sRL}{R+sL}}$

(b) $i(t) = 10e^{-3000t}\cos(2000t + 20°)\;\text{mA}$

$\mathbf{V}_{op} = \frac{(-3000 + j2000)(1)(1\times10^{-3})}{1 + (-3000 + j2000)(1\times10^{-3})}(10\angle20°\;\text{mA}) = 12.7\angle31.31°\;\text{mV}$

$\boxed{v_{op}(t) = 12.7e^{-3000t}\cos(2000t + 31.31°)\;\text{mV}}$

12. $s = j1000\;\text{rad/s}$

$$\mathbf{V}_{op} = \frac{j1000(1)(1\times10^{-3})}{1 + j1000(1\times10^{-3})}(10\angle20° \text{ mA}) = 7.071\angle65° \text{ mV}$$

$$\boxed{v_{op}(t) = 7.071\cos(1000t + 65°) \text{ mV}}$$

13. $s = -3000$ n/s

$$\mathbf{V}_{op} = \frac{-3000(1)(1\times10^{-3})}{1 - 3000(1\times10^{-3})}(10\angle0° \text{ mA}) = 15\angle0° \text{ mV}$$

$$\boxed{v_{op}(t) = 15e^{-3000t} \text{ mV}}$$

14. $\mathcal{A}(p)v_{op}(t) = \mathcal{B}(p)v_s(t)$

Let $v_s(t) = \mathbf{V}_s e^{st}$: then, $v_{op}(t) = \mathbf{V}_{op} e^{st}$.

$\mathcal{A}(p)\mathbf{V}_{op} e^{st} = \mathcal{B}(p)\mathbf{V}_s e^{st}$

Perform the differentiations to yield

$\mathcal{A}(s)\mathbf{V}_{op} e^{st} = \mathcal{B}(s)\mathbf{V}_s e^{st} \Rightarrow \mathbf{V}_o = \mathbf{H}(s)\mathbf{V}_s$

15.   Resistance

$v(t) = Ri(t)$

If $i(t) = \mathbf{I}e^{st}$, then $v(t) = \mathbf{V}e^{st}$.

$\mathbf{V} = R\mathbf{I}$

Capacitance

$i(t) = C\dfrac{d}{dt}v(t)$

If $i(t) = \mathbf{I}e^{st}$, then $v(t) = \mathbf{V}e^{st}$.

$\mathbf{I} = sC\mathbf{V}$

$\mathbf{V} = \dfrac{1}{sC}\mathbf{I}$

Inductance

$v(t) = L\dfrac{d}{dt}i(t)$

If $i(t) = \mathbf{I}e^{st}$, then $v(t) = \mathbf{V}e^{st}$.

$\mathbf{V} = sL\mathbf{I}$

Mutual Inductance

$v_1 = L_1\dfrac{d}{dt}i_1 + M\dfrac{d}{dt}i_2$

$v_2 = M\dfrac{d}{dt}i_1 + L_2\dfrac{d}{dt}i_2$

$\mathbf{V}_1 = sL_1\mathbf{I}_1 + sM\mathbf{I}_2$

$\mathbf{V}_2 = sM\mathbf{I}_1 + sL_1\mathbf{I}_2$

16.   (a)

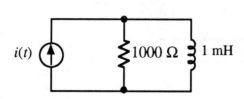

        (b)

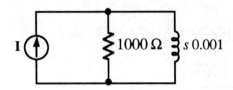

(c) $\quad \mathbf{Z}(s) = R \parallel sL = \dfrac{sRL}{R+sL}$

(d) $\quad \mathbf{Z}(j10^6) = \dfrac{1000}{\sqrt{2}} \angle 45° = 707.11 \angle 45° \; \Omega$

$\Rightarrow \boxed{v_{op}(t) = 3535.5 \cos(10^6 + 45°) \; \text{V}}$

17. $\quad \sum i_n(t) = \sum I_{mn} e^{\sigma t} \cos(\omega t + \phi_n) = \sum \mathcal{R}e\{\mathbf{I}_n e^{st}\} = \mathcal{R}e\{\sum \mathbf{I}_n e^{st}\}$

The equation $\sum \mathbf{I}_n = 0$ is therefore a sufficient conditon for $\sum i_n(t) = 0$.

18. Given the circuit shown below

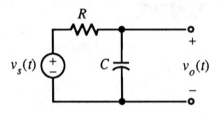

(a) Voltage divider: $H(s) = \dfrac{1}{1+s}$

(b)

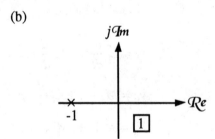

(c) $\quad s_1 = -1$

(d) $\quad v_{oc}(t) = Ae^{-t}$

(e) $\quad s = j \Rightarrow \boxed{v_{op}(t) = 7.07 \cos(t - 25°) \; \text{V}}$

19.   Given the circuit shown below:

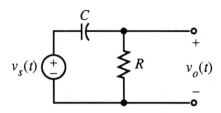

(a)   Voltage divider: $\mathbf{H}(s) = \dfrac{s}{1+s}$

(b)

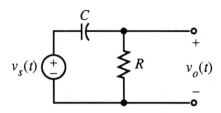

(c) $s_1 = -1$

(d) $v_{oc}(t) = Ae^{-t}$

(e) $v_{op}(t) = 7.07\cos(t + 65°)\ \text{V}$

20.   (a) & (b) By substituting $\mathbf{I}e^{st}$ and $\mathbf{V}e^{st}$ into the differential equation and solving for $\mathbf{V}$, we obtained $\mathbf{V} = \mathbf{Z}(s)\mathbf{I}$, where

$$\boxed{\mathbf{Z}(s) = \dfrac{s(s+7)}{(s+10)(s+5)}}$$

(c)

(d) $\mathbf{Z}(j5) = \dfrac{j5(j5+7)}{(j5+10)(j5+5)} = 0.544\angle 54°$

(e) $\mathbf{V} = 10\angle 0°(0.544\angle 54°) = 5.44\angle 54°$

(f) $v_p(t) = 5.44\cos(5t + 54°)$

21.   (a) & (b) By substituting $\mathbf{I}e^{st}$ and $\mathbf{V}e^{st}$ into the differential equation and solving for $\mathbf{V}$, we obtained $\mathbf{V} = \mathbf{Z}(s)\mathbf{I}$, where

$$\boxed{\mathbf{Z}(s) = \frac{(s+3+j5)(s+3-j5)}{(s+10)(s+5)}}$$

(c)

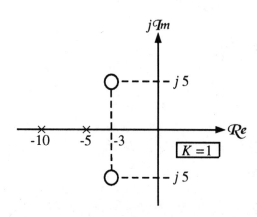

(d) $\mathbf{Z}(j5) = \dfrac{(j5+3+j5)(j5+3-j5)}{(j5+10)(j5+5)} = 0.396\angle 1.73°$

(e) $\mathbf{V} = 10\angle 0°(0.396\angle 1.73°) = 3.96\angle 1.73°$

(f) $v_p(t) = 3.96\cos(5t+1.73°)$

22.   An approximation of the magnitude response is shown below.

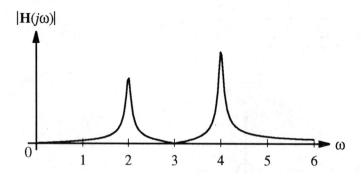

23.   Given the circuit shown below:

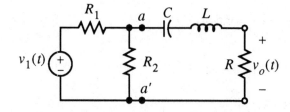

(a)   Replace everything to the left of terminals *a-a'* with its Thévenin equivalent circuit as shown below.

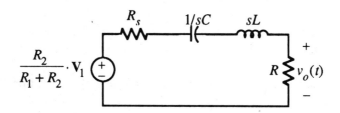

(b)   Voltage divider: $\dfrac{\mathbf{V}_o}{\mathbf{V}_1} = \mathbf{H}(s) = \dfrac{R}{R + R_s + \frac{1}{sC} + sL}\left(\dfrac{R_2}{R_1 + R_2}\right)$

$$\mathbf{H}(s) = \dfrac{s\left[\dfrac{RR_2}{(R_1 + R_2)L}\right]}{s^2 + s\left(\dfrac{R + R_s}{L}\right) + \dfrac{1}{LC}} \quad \text{where } R_s = R_1 \parallel R_2$$

(c)   The coefficient of s is in the denominator gives the bandwidth.

$$\Rightarrow \quad BW = \dfrac{R + R_s}{L}$$

24.   Given the circuit shown below:

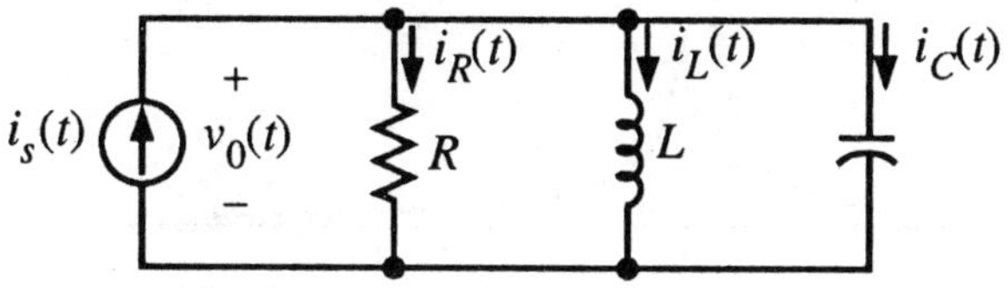

(a)   From current divider: $\mathbf{I}_R = \dfrac{sL \parallel \frac{1}{sC}}{R + sL \parallel \frac{1}{sC}}\mathbf{I}_s$

$$\Rightarrow \quad \mathbf{H}(s) = \dfrac{s\left(\frac{1}{RC}\right)}{s^2 + s\left(\frac{1}{RC}\right) + \frac{1}{LC}}$$

(b)   The dual of the circuit is shown below.

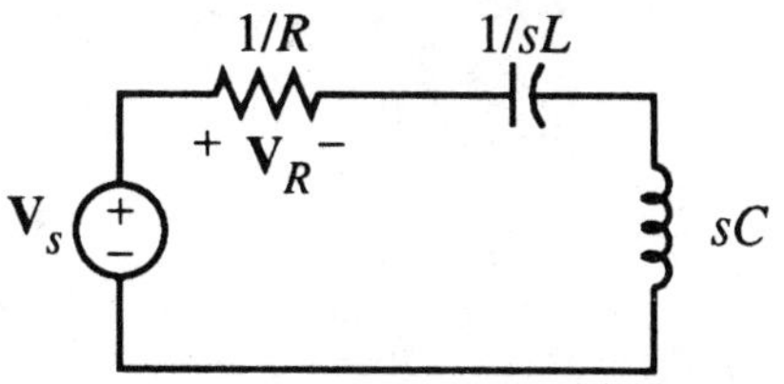

By voltage divider: $\dfrac{\mathbf{V}_R}{\mathbf{V}_s} = \dfrac{\frac{1}{R}}{\frac{1}{R} + sC + \frac{1}{sL}} = \dfrac{s\left(\frac{1}{RC}\right)}{s^2 + s\left(\frac{1}{RC}\right) + \frac{1}{LC}}$

(c) $\Rightarrow \omega_o = \dfrac{1}{\sqrt{LC}}$

(d) $\Rightarrow BW = \dfrac{1}{RC}$

25.   Using the results from Excercise 24 with $C = 1\ \mu F$ and $L = 2.5$ mH,

(a) $BW = 1000 = \dfrac{1}{R(1\ \mu F)} \quad \Rightarrow \quad \boxed{R = 1000\ \Omega}$

(b) $H(s) = \dfrac{1000s}{s^2 + 1000s + 4 \times 10^8}$

$|H(j\omega)| = \dfrac{10^3 \omega}{\sqrt{\left(4 \times 10^8 - \omega^2\right)^2 + 10^6 \omega^2}}$

$\angle H(j\omega) = 90° - \text{arc tan}\left[\dfrac{10^3 \omega}{4 \times 10^8 - \omega^2}\right]$

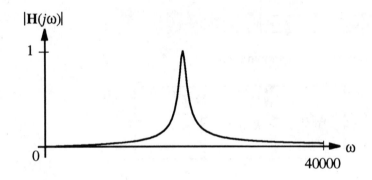

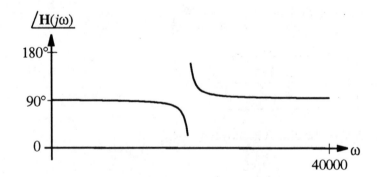

26. $Y(j\omega) = \dfrac{1}{R} + j\left(\omega C - \dfrac{1}{\omega L}\right)$

Set $\mathcal{I}m\{Y(j\omega)\} = 0$ to obtain $\boxed{\omega_{upf} = \omega_o = \dfrac{1}{\sqrt{LC}}}$

Using duality, all we have to do is interchange the L and the C since the resistor plays no part in this calculation. The result is the same $\Rightarrow \omega_{upf} = \omega_o = \dfrac{1}{\sqrt{LC}}.$

27.   Since the susceptance is zero for ω_{upf}, no current can enter the closed part of the circuit. The reactance looks like an open circuit (parallel resonance). Therefore, $\mathbf{I}_R = \mathbf{I}_s$ and $\mathbf{V}_o = \mathbf{I}_s R$. Then $\mathbf{I}_c = j\omega_o RC\mathbf{I}_s$ and $\mathbf{I}_L = -j\omega_o RC\mathbf{I}_s$.

28.   For series *RLC* circuit, $Q = \dfrac{\omega_o}{\text{BW}}$, where $\text{BW} = \dfrac{R}{L}$.

(a) $\Rightarrow$ $\boxed{\text{BW} = \dfrac{200,000\pi}{10} = 20,000\pi \text{ rad/s}}$

(b) $P_{av} = \dfrac{V_{rms}^2}{R} = 2 \text{ mW} \Rightarrow \boxed{R = 8000 \ \Omega}$

$20,000\pi = \dfrac{R}{L} \Rightarrow \boxed{L = 0.127 \text{ H}} \ : \ \dfrac{1}{\sqrt{LC}} = 200,000\pi \Rightarrow \boxed{C = 20 \text{ pF}}$

29.   For a parallel *RLC* circuit, $Q = \dfrac{\omega_o}{\text{BW}}$, where $\text{BW} = \dfrac{1}{RC}$.

(a) $\Rightarrow$ $\boxed{\text{BW} = \dfrac{2000\pi}{100} = 20\pi \text{ rad/s}}$

(b) $P_{av} = I_{rms}^2 R = 1 \text{ W} \Rightarrow \boxed{R = 1 \ \Omega}$

(c) $\dfrac{1}{RC} = 2000\pi \Rightarrow \boxed{C = 15.9 \text{ mF}} \ : \ 2000\pi = \dfrac{1}{\sqrt{LC}} \Rightarrow \boxed{L = 1.59 \ \mu\text{H}}$

30. $BW = 2\alpha = 4000 \Rightarrow \alpha = 2000$

$\gamma = \dfrac{\omega - \omega_d}{\alpha} = \dfrac{6000 - 1000}{2000} = \dfrac{5}{2}$

$|\mathbf{H}(j\omega)| = \dfrac{1}{\sqrt{1+\gamma^2}}|\mathbf{H}(j\omega_d)| = \dfrac{1}{\sqrt{1+\dfrac{25}{4}}}(20) \Rightarrow \boxed{|\mathbf{H}(j\omega)| = 7.43}$

31. $BW = 2\alpha = 4000 \Rightarrow \alpha = 2000$

$\gamma = \dfrac{\omega - \omega_d}{\alpha} = \dfrac{8000 - 1000}{2000} = \dfrac{7}{2}$

$|\mathbf{H}(j\omega)| = \dfrac{1}{\sqrt{1+\gamma^2}}|\mathbf{H}(j\omega_d)| = \dfrac{1}{\sqrt{1+\dfrac{49}{4}}}(20) \Rightarrow \boxed{|\mathbf{H}(j\omega)| = 5.49}$

32. $\mathbf{H}_c(j\omega) = \dfrac{1}{1 - \omega^2 LC + j\omega RC}$

It can be shown that $\omega_{mr} = \sqrt{\dfrac{1}{LC} - 2\left(\dfrac{R}{2L}\right)^2}$, if $\dfrac{1}{LC} - 2\left(\dfrac{R}{2L}\right)^2 > 0$

$$\omega_{mr} = \sqrt{\omega_o^2 - 2\left(\frac{1}{2}\frac{\omega_o}{Q}\right)^2} = \omega_o\sqrt{1 - \frac{1}{2Q^2}} = \omega_o\sqrt{1 - 2\zeta^2}$$

It has been described in the text that $\mathbf{H}_R$, $\omega_{mr} = \omega_o = \dfrac{1}{\sqrt{LC}}$.

$$\mathbf{H}_L = \frac{-\omega^2 LC}{1 - \omega^2 LC + j\omega RC}$$

It can be shown that from $\dfrac{d}{d\omega}\left|\mathbf{H}_L(j\omega)\right| = 0$ that $\omega_{mr} = \dfrac{1/LC}{\sqrt{\dfrac{1}{LC} - \dfrac{1}{2}\left(\dfrac{R}{L}\right)^2}}$,

if $\dfrac{1}{LC} - \dfrac{1}{2}\left(\dfrac{R}{L}\right)^2 > 0$

$$\omega_{mr} = \frac{1/LC}{\sqrt{\dfrac{1}{LC} - \dfrac{1}{2}\left(\dfrac{R}{L}\right)^2}} = \frac{\omega_o^2}{\sqrt{\omega_o^2 - \dfrac{1}{2}(2\alpha)^2}} = \frac{\omega_o}{\dfrac{\sqrt{\omega_o^2 - \dfrac{1}{2}(2\alpha)^2}}{\omega_o}}$$

$$\omega_{mr} = \frac{\omega_o}{\sqrt{1 - \dfrac{1}{2}(2\zeta)^2}} = \frac{\omega_o}{\sqrt{1 - 2\zeta^2}}$$

33.   (a) $\mathbf{H}_C(0) = 1$ since the capacitor looks like an open circuit at dc and the inductor looks like a short circuit.

         (b) $\mathbf{H}_L(\infty) = 1$ since the inductor looks like an open circuit and the capacitor looks like a short circuit.

         (c) $\mathbf{H}_R(0) = \mathbf{H}_R(\infty) = 0$ since either the capacitor or the inductor is a short circuit or an open circuit at dc and at ∞.

34.   Given the circuit shown below:

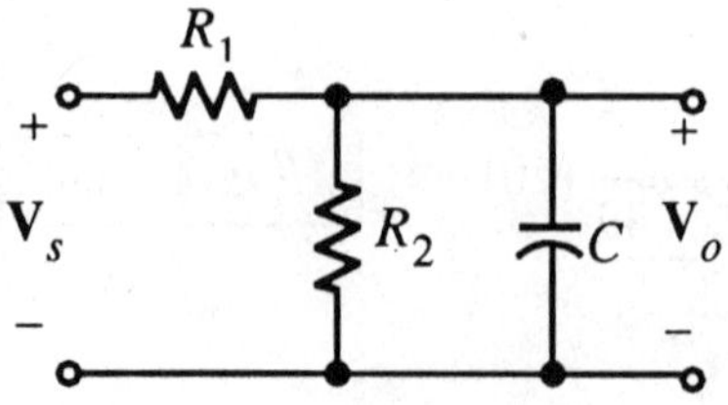

$$\mathbf{H}(s) = \frac{R_2}{R_1 + R_2 + sR_1R_2C} = \frac{0.1}{s + 0.2} = \frac{0.5}{1 + s/0.2}$$

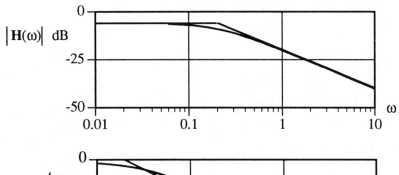

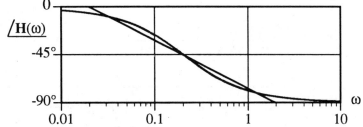

35.   Given the circuit shown below:

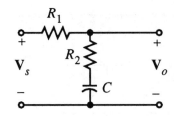

$$H(s) = \frac{1+sR_2C}{1+s(R_1+R_2)C} = \frac{1+10s}{1+20s}$$

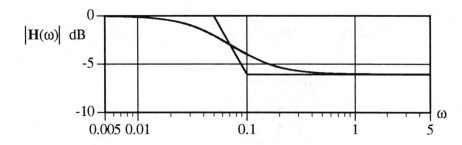

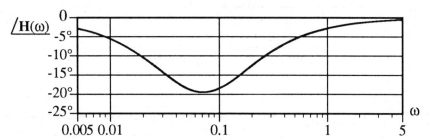

36.   Given the circuit shown below:

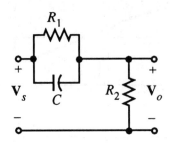

$$H(s) = \frac{R_2 + sR_1R_2C}{R_1 + R_2 + sR_1R_2C} = \frac{1}{2} \cdot \frac{1+10s}{1+5s}$$

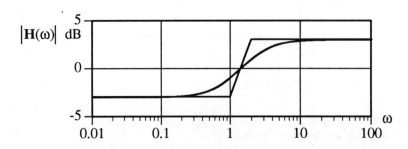

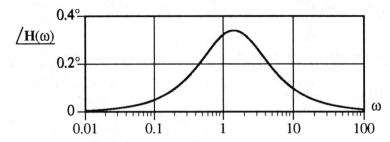

37.   Given the circuit shown below:

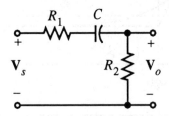

$$H(s) = \frac{sR_2C}{1 + s(R_1 + R_2)C} = \frac{s}{0.1 + 2s}$$

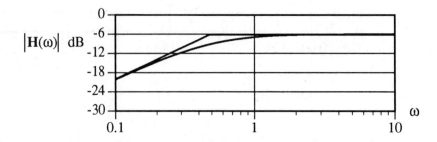

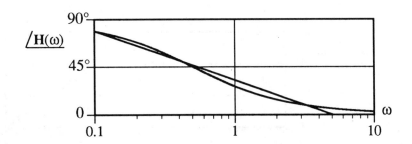

38.   Given the circuit shown below:

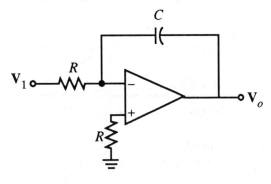

$$H(s) = \frac{\mathbf{V}_o}{\mathbf{V}_1} = -\frac{1}{sRC} = -\frac{1000}{s}$$

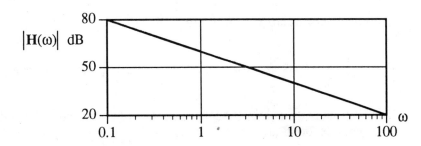

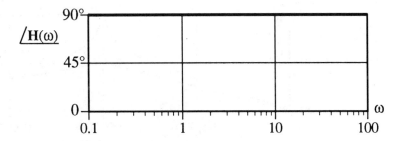

39.   For the given series *RLC* circuit with $R = 0.1\,\Omega$, $L = 1\,\mu H$, and $C = 1\,\mu F$, if the output is the voltage across the capacitor, its transfer function is given by

$$H(s) = \frac{1}{s^2 LC + sRC + 1} = \frac{1}{10^{-12} s^2 + 10^{-7} s + 1}$$

The pole locations are $s = -50,000 \pm j998,749.22$

$$\zeta = \frac{10^{-7}}{2(10^{-6})} = 0.05 \implies \text{roughly 20 dB peak above the break frequency.}$$

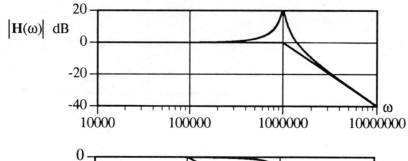

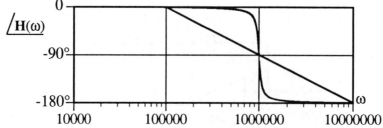

40. For the given series *RLC* circuit with $R = 0.1\,\Omega$, $L = 1\,\mu H$, and $C = 1\,\mu F$, if the output is the voltage across the resistor, its transfer function is given by

$$H(s) = \frac{s(R/L)}{s^2 + s(R/L) + 1/LC} = \frac{10^5 s}{s^2 + 10^5 s + 10^{12}} = \frac{10^{-7} s}{10^{-12} s^2 + 10^{-7} s + 1}$$

The pole locations are $s = -50,000 \pm j998,749.22$

$$\zeta = \frac{10^{-7}}{2(10^{-6})} = 0.05 \quad \Rightarrow \quad \text{roughly 20 dB peak above the break frequency.}$$

The "s" term in the numerator contirbutes 90° of phase and a 20 dB/decade shift in magnitude for all ω.

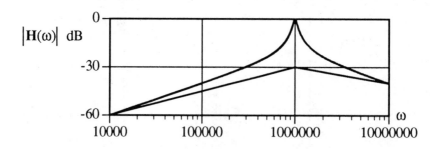

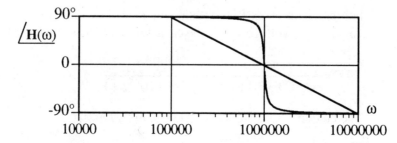

41. For the given series *RLC* circuit with $R = 0.1\,\Omega$, $L = 1\,\mu H$, and $C = 1\,\mu F$, if the output is the voltage across the inductor, its transfer function is given by

$$H(s) = \frac{s^2 LC}{s^2 LC + sRC + 1} = \frac{10^{-12} s^2}{10^{-12} s^2 + 10^{-7} s + 1}$$

The pole locations are $s = -50,000 \pm j998,749.22$

$$\zeta = \frac{10^{-7}}{2(10^{-6})} = 0.05 \implies \text{roughly 20 dB peak above the break frequency.}$$

The "s^2" term in the numerator contributes 180° of phase and and a 40 dB/decade shift in magnitude for all ω.

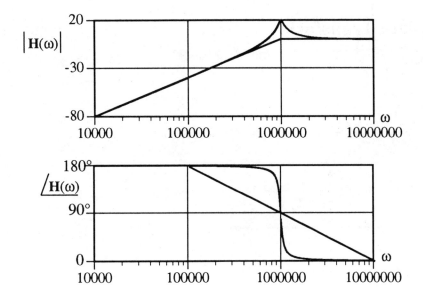

CHAPTER 14

The Laplace Transform

Exercises

1.
$$v(t) = u(t-1), \implies V(s) = \frac{e^{-s}}{s}, \mathcal{R}e\{s\} > 0$$

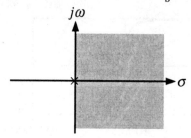

2.
$$v(t) = u(t) - u(t-1), \implies V(s) = \frac{1}{s} - \frac{e^{-s}}{s}, \mathcal{R}e\{s\} > 0$$

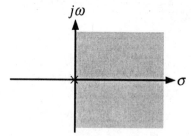

3.
$$v(t) = e^{-3t}u(t) + e^{-5t}u(t), \implies V(s) = \frac{1}{s+3} + \frac{1}{s+5}, \mathcal{R}e\{s\} > -3$$

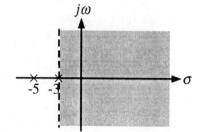

4. $$v(t) = e^{-3t}u(t) + e^{5t}u(t), \Rightarrow V(s) = \frac{1}{s+3} + \frac{1}{s-5}, \mathcal{R}e\{s\} > 5$$

5. By setting $\alpha = -2$ in Example 14.1 we have $v(t) = e^{-2t}u(t)$.

6. $$V(s) = 1 + \frac{1}{s} + \frac{1}{s+2} \Rightarrow v(t) = \mathcal{L}^{-1}[V(s)] = \mathcal{L}^{-1}\left[1 + \frac{1}{s} + \frac{1}{s+2}\right]$$

    Since $\mathcal{L}^{-1}$ is a linear operator, this becomes

    $$\boxed{v(t) = \mathcal{L}^{-1}(1) + \mathcal{L}^{-1}\left(\frac{1}{s}\right) + \mathcal{L}^{-1}\left(\frac{1}{s+2}\right) = \delta(t) + u(t) + e^{-2t}u(t)}$$

7. $$\mathcal{L}[tu(t)] = \frac{1}{s^2}$$

8. $$\mathcal{L}\left[(t-t_0)u(t-t_0)\right] = \frac{e^{-st_0}}{s^2} \quad \text{using } LT3 \text{ and the delay theorem.}$$

9. Note that $(t-t_0)u(t-t_0) = tu(t-t_0) - t_0 u(t-t_0)$

    $$\Rightarrow tu(t-t_0) = (t-t_0)u(t-t_0) + t_0 u(t-t_0)$$

    $$\mathcal{L}[tu(t-t_0)] = \mathcal{L}\left[(t-t_0)u(t-t_0) + t_0 u(t-t_0)\right] = \frac{e^{-st_0}}{s^2} + t_0\frac{e^{-st_0}}{s}$$

    $$\boxed{\mathcal{L}[tu(t-t_0)] = \frac{e^{-st_0}}{s^2} + t_0\frac{e^{-st_0}}{s} = \frac{1+st_0}{s^2}e^{-st_0}}$$

10. $$f(t) = 5te^{7t}u(t) + 2\cos(\omega_0 t + 45°)u(t)$$

    $$\mathcal{L}[f(t)] = \frac{5}{(s-7)^2} + \frac{(1/2)(2\angle 45°)}{s-j\omega_0} + \frac{(1/2)(2\angle -45°)}{s+j\omega_0}$$

    $$\Rightarrow \boxed{\mathcal{L}[f(t)] = \frac{5}{(s-7)^2} + \frac{1\angle 45°}{s-j\omega_0} + \frac{1\angle -45°}{s+j\omega_0}}$$

11. $$\Pi\left(\frac{t-0.5\tau}{\tau}\right) = u(t) - u(t-\tau)$$

    $$\mathcal{L}\left[\Pi\left(\frac{t-0.5\tau}{\tau}\right)\right] = \mathcal{L}[u(t) - u(t-\tau)] \Rightarrow \boxed{\mathcal{L}\left[\Pi\left(\frac{t-0.5\tau}{\tau}\right)\right] = \frac{1}{s} - \frac{e^{-s\tau}}{s}}$$

12. $$f(t) = \mathcal{L}^{-1}\left(\frac{e^{-s}}{s}\right) \Rightarrow \boxed{f(t) = u(t-1)}$$

13.
$$V(s) = \frac{1}{s - j30 + 10} \quad \Rightarrow \quad \boxed{v(t) = e^{(-10+j30)t}u(t)}$$

14.
$$V(s) = \frac{s-2}{s^2 - 2s + 2} \quad \Rightarrow \quad \frac{s-2}{(s-1)^2 + 1} = \frac{s-1}{(s-1)^2 + 1} - \frac{1}{(s-1)^2 + 1}$$

$$\Rightarrow \quad \boxed{v(t) = e^t \left[\cos(t) - \sin(t)\right]u(t)}$$

15.
$$V(s) = \frac{s+1}{(s+2)(s+3)} = \frac{a}{s+2} + \frac{b}{s+3} \quad \Rightarrow \quad s+1 = a(s+3) + b(s+1)$$

$$\Rightarrow \quad a = -1 \text{ and } b = 2 \quad \Rightarrow \quad V(s) = \frac{s+1}{(s+2)(s+3)} = \frac{-1}{s+2} + \frac{2}{s+3}$$

$$\Rightarrow \quad \boxed{v(t) = -e^{-2t}u(t) + 2e^{-3t}u(t)}$$

16. Set $\alpha = 0$ in pair $LT11$.

17. Set $\alpha = 0$ in pair $LT12$.

18. Let $K_o = 1\angle 0° = 1$

$$e^{\sigma_0 t}\cos(\omega_0 t) \Leftrightarrow \frac{1/2}{s - s_0} + \frac{1/2}{s - s_0^*} = \frac{1}{2}\left(\frac{s - s_0^* + s - s_0}{s^2 - ss_0^* - ss_0 + s_0 s_0^*}\right)$$

$$\Rightarrow e^{\sigma_0 t}\cos(\omega_0 t) \Leftrightarrow \frac{1}{2}\left[\frac{2s - 2\sigma_0}{(s+\sigma_0)^2 + \omega_0}\right] = \frac{s - \sigma_0}{(s+\sigma_0)^2 + \omega_0^2}$$

Let $\sigma_0 = -\alpha$; then

$$\boxed{e^{\alpha t}\cos(\omega_0 t) \Leftrightarrow \frac{s+\alpha}{(s+\alpha)^2 + \omega_0^2}}$$

19. Set $K_0 = j$ and $\sigma_0 = -\alpha$.

Then $LT9$ becomes

$$e^{-\alpha t}\cos(\omega_0 t + 90°) \Leftrightarrow \frac{j/2}{s - (-\alpha + j\omega_0)} - \frac{j/2}{s - (-\alpha - j\omega_0)}$$

$$\Rightarrow \quad \boxed{e^{-\alpha t}\sin(\omega_0 t) \Leftrightarrow \frac{\omega_0}{(s+\alpha)^2 + \omega_0^2}}$$

20. Initial value: $v(0^+) = \lim_{\mathcal{R}e\{s\} \to \infty} sV(s) = \lim_{\mathcal{R}e\{s\} \to \infty} s\left(\frac{e^{-3s}}{s}\right) = 0$

Final value: $\lim\limits_{\substack{\mathcal{R}e\{s\} \to 0}} sV(s) = \lim\limits_{\substack{\mathcal{R}e\{s\} \to 0}} s\left(\dfrac{e^{-3s}}{s}\right) = 1$

21.

Initial value: $v(0^+) = \lim\limits_{\substack{\mathcal{R}e\{s\} \to \infty}} sV(s) = \lim\limits_{\substack{\mathcal{R}e\{s\} \to \infty}} \left(\dfrac{s^2}{(s+1)(s+2)}\right) = 1$

Final value: $\lim\limits_{\substack{\mathcal{R}e\{s\} \to 0}} sV(s) = \lim\limits_{\substack{\mathcal{R}e\{s\} \to 0}} \left(\dfrac{s^2}{(s+1)(s+2)}\right) = 0$

22.

Initial value: $v(0^+) = \lim\limits_{\substack{\mathcal{R}e\{s\} \to \infty}} sV(s) = \lim\limits_{\substack{\mathcal{R}e\{s\} \to \infty}} \dfrac{s(s+3)e^{-s}}{s^2 + 6s + 18} = 0$

Final value: $\lim\limits_{\substack{\mathcal{R}e\{s\} \to 0}} sV(s) = \lim\limits_{\substack{\mathcal{R}e\{s\} \to 0}} \dfrac{s(s+3)e^{-s}}{s^2 + 6s + 18} = 0$

23.

$\dfrac{1}{s^2 + 1} = \dfrac{a}{s+j} + \dfrac{b}{s-j} \quad : a = \dfrac{1}{s-j}\bigg|_{s=-j} = \dfrac{1}{-j2} = j0.5 \ : \ b = a^* = -j0.5$

$$\boxed{\dfrac{1}{s^2 + 1} = \dfrac{j0.5}{s+j} - \dfrac{j0.5}{s-j}}$$

24.

$\dfrac{s}{(s+1)^2} = \dfrac{a}{s+1} + \dfrac{b}{(s+1)^2} \quad : b = (s+1)^2 \dfrac{s}{(s+1)^2}\bigg|_{s=-1} = -1$

$\dfrac{s}{(s+1)^2} = \dfrac{a}{s+1} - \dfrac{1}{(s+1)^2} \ \Rightarrow \ s = a(s+1) - 1 \ \Rightarrow \ a = 1$

$$\boxed{\dfrac{s}{(s+1)^2} = \dfrac{1}{s+1} - \dfrac{1}{(s+1)^2}}$$

25.

$\dfrac{s^2 + 2s + 2}{s^2 + 7s + 12} = 1 - \dfrac{5s + 10}{s^2 + 7s + 12}$ (by long division)

$\dfrac{5s + 10}{s^2 + 7s + 12} = \dfrac{a}{s+3} + \dfrac{b}{s+4} : a = \dfrac{5s+10}{s+4}\bigg|_{s=-3} = -5 \ : \ b = \dfrac{5s+10}{s+3}\bigg|_{s=-4} = 10$

$$\boxed{\dfrac{s^2 + 2s + 2}{s^2 + 7s + 12} = 1 + \dfrac{5}{s+3} - \dfrac{10}{s+4}}$$

26.

$\dfrac{s^3 + 3s^2 + 2s + 1}{s+2} = s^2 + s + \dfrac{1}{s+2}$ (by long division)

27.

$\dfrac{(s^3 + 3s^2 + 2s + 1)(s^2 + 2s + 2)}{(s+2)(s^2 + 7s + 12)} = \dfrac{s^5 + 5s^4 + 10s^3 + 11s^2 + 6s + 2}{s^3 + 9s^2 + 26s + 24}$

By long division this becomes

$$\frac{\left(s^3+3s^2+2s+1\right)\left(s^2+2s+2\right)}{(s+2)\left(s^2+7s+12\right)}=s^2-4s+20-\frac{89s^2+514s+480}{(s+2)\left(s^2+7s+12\right)}$$

$$\frac{89s^2+514s+480}{(s+2)\left(s^2+7s+12\right)}=\frac{a}{s+2}+\frac{b}{s+3}+\frac{c}{s+4}$$

$a=239,\ b=-923,\ c=259$

$$\boxed{\frac{\left(s^3+3s^2+2s+1\right)\left(s^2+2s+2\right)}{(s+2)\left(s^2+7s+12\right)}=s^2-4s+20-\frac{239}{s+2}+\frac{923}{s+3}-\frac{259}{s+4}}$$

28.
$$\frac{s\left[(s+3)^2+1\right]}{(s+5)\left[(s+1)^2+9\right]}=\frac{s^3+6s^2+10s}{s^3+7s^2+20s+50}=1-\frac{s^2+10s+50}{s^3+7s^2+20s+50}$$

$$\frac{s^2+10s+50}{s^3+7s^2+20s+50}=\frac{a}{s+5}+\frac{bs+c}{s^2+2s+10}\ :\ a=1$$

$$\Rightarrow\ s^2+10s+50=1\left(s^2+2s+10\right)+(bs+c)(s+5)$$

$$\Rightarrow\ s^2=(1+b)s^2\ \Rightarrow\ b=0\ :\ 50=10+5c\ \Rightarrow\ c=8$$

$$\frac{s^2+10s+50}{s^3+7s^2+20s+50}=\frac{1}{s+5}+\frac{8}{s^2+2s+10}$$

$$\boxed{\frac{s\left[(s+3)^2+1\right]}{(s+5)\left[(s+1)^2+9\right]}=1-\frac{1}{s+5}-\frac{8}{s^2+2s+10}=1-\frac{1}{s+5}-\frac{8}{(s+1)^2+9}}$$

29.
$$G_a(s)=\frac{k_a}{s-s_a},\ G_b(s)=\frac{k_b}{s-s_b}$$

$$G_a(s)G_b(s)=\frac{k_ak_b}{\left(s-s_a\right)\left(s-s_b\right)}=\frac{a}{s-s_a}+\frac{b}{s-s_b}\ :\ a=\frac{k_ak_b}{s_a-s_b},b=\frac{k_ak_b}{s_b-s_a}=-a$$

$$\boxed{G_a(s)G_b(s)=\frac{k_ak_b}{\left(s-s_a\right)\left(s-s_b\right)}=\frac{a}{s-s_a}+\frac{a}{s-s_b}\ \text{where}\ a=\frac{k_ak_b}{s_a-s_b}}$$

30.
$$G_a(s)=\frac{k_a}{s-s_a},\ G_b(s)=\frac{k_b}{s-s_a}$$

$$G_a(s)G_b(s) = \frac{k_a k_b}{(s-s_a)^2} = \frac{a}{s-s_a} + \frac{b}{(s-s_b)^2}$$

$$\Rightarrow k_a k_b = a(s-s_a) + b \Rightarrow a = 0, b = k_a k_b$$

$$\boxed{G_a(s)G_b(s) = \frac{k_a k_b}{(s-s_a)^2}}$$

31.
$$\frac{s}{s^2 + \omega_o^2} = \frac{a}{s-j\omega_o} + \frac{a^*}{s+j\omega_o} \quad \text{where } a = \frac{j\omega_o}{j2\omega_o} = \frac{1}{2}$$

32.
$$\frac{\omega_o}{s^2 + \omega_o^2} = \frac{a}{s-j\omega_o} + \frac{a^*}{s+j\omega_o} \quad \text{where } a = \frac{\omega_o}{j2\omega_o} = \frac{1}{j2} = -j0.5$$

33.
$$\frac{s+\alpha}{(s+\alpha-j\omega_o)(s+\alpha+j\omega_o)} = \frac{a}{s+\alpha-j\omega_o} + \frac{a^*}{s+\alpha+j\omega_o} \quad \text{where } a = \frac{j\omega_o}{j2\omega_o} = \frac{1}{2}$$

34.
$$\frac{\omega_o}{(s+\alpha-j\omega_o)(s+\alpha+j\omega_o)} = \frac{a}{s+\alpha-j\omega_o} + \frac{a^*}{s+\alpha+j\omega_o}$$

$$\text{where } a = \frac{\omega_o}{j2\omega_o} = \frac{1}{j2} = -j0.5$$

35. Given the circuit shown below:

$$\frac{1}{C}\int_{-\infty}^{t} i(\lambda)d\lambda + v_R(t) = v_s(t) = u(t)$$

$$\frac{1}{s} = \frac{I(s)}{s} + \frac{v_C(0^-)}{s} + V_R(s)$$

$$\frac{1}{s} = \frac{V_R(s)}{s} + V_R(s) = V_R(s)\left(1+\frac{1}{s}\right) \Rightarrow V_R(s) = \frac{1}{s+1} \Rightarrow \boxed{v_R(t) = e^{-t}u(t) \text{ V}}$$

36. Refer to Exercise 35 with $v_s(t) = 10tu(t)$ and $v_C(0^-) = 6$.

$$\frac{10}{s^2} = \frac{V_R(s)}{s} + \frac{6}{s} + V_R(s) \Rightarrow V_R(s) = \frac{s}{s+1} \cdot \frac{10-6s}{s^2} = \frac{10-6s}{s(s+1)} = \frac{a}{s} + \frac{b}{s+1}$$

$$a = 10, \ b = -16$$

$$\Rightarrow \boxed{v_R(t) = (10 - 16e^{-t})u(t) \text{ V}}$$

37.   Given the circuit below:

NOTE: $v_R(0^-) = Ri_L(0^-)$

$$v_s(t) = Ri_L(t) + v_L(t), \text{ where } i_L(t) = \frac{1}{L}\int_{-\infty}^{t} v_L(\lambda)d\lambda$$

$$\frac{R}{L}\int_{-\infty}^{t} v_L(\lambda)d\lambda + v_L(t) = v_s(t)$$

$$V_s(s) = \frac{1}{s} = \frac{R}{sL}V_L(s) + \frac{Ri_L(0^-)}{s} + V_L(s) = V_L(s)\left(\frac{R+sL}{sL}\right) + \frac{v_R(0^-)}{s}$$

$$\Rightarrow V_L(s) = \frac{L[1 - v_R(0^-)]}{R+sL} = \frac{1 - v_R(0^-)}{s + R/L} \Rightarrow \boxed{v_L(t) = [1 - v_R(0^-)]e^{-\frac{R}{L}t}u(t)}$$

38.   Refer to Exercise 37 above

$$V_s(s) = V_L(s)\left(\frac{R+sL}{sL}\right) + \frac{v_R(0^-)}{s} = \frac{1/2(A\angle\phi)}{s - j\omega_o} + \frac{1/2(A\angle-\phi)}{s + j\omega_o}$$

$$V_L(s) = \frac{1/2(A\angle-\phi)s}{(s - j\omega_o)(s + R/L)} + \frac{1/2(A\angle-\phi)s}{(s + j\omega_o)(s + R/L)} - \frac{v_R(0^-)}{s + R/L}$$

$$\frac{1/2(A\angle\phi)s}{(s - j\omega_o)(s + R/L)} = \frac{a_1}{s - j\omega_o} + \frac{b_1}{s + R/L}$$

$$a_1 = \frac{1/2(A\angle\phi)j\omega_o}{j\omega_o + R/L} = \frac{1/2(A\omega_o)}{\sqrt{(R/L)^2 + \omega_o^2}}\angle\phi + 90° - \tan^{-1}\left(\frac{\omega_o L}{R}\right) = K_o\angle\theta_o$$

$$b_1 = \frac{1/2(A\angle\phi)(-R/L)}{-R/L - j\omega_o} = \frac{1/2(AR/L)}{\sqrt{(R/L)^2 + \omega_o^2}}\angle\phi - \tan^{-1}\left(\frac{\omega_o L}{R}\right) = K_1\angle\theta_1$$

$$\frac{1/2(A\angle-\phi)s}{(s + j\omega_o)(s + R/L)} = \frac{a_2}{s + j\omega_o} + \frac{b_2}{s + R/L}$$

$$a_2 = \frac{1/2(A\angle-\phi)(-j\omega_o)}{(-j\omega_o + R/L)} = \frac{1/2(A\omega_o)}{\sqrt{(R/L)^2 + \omega_o^2}}\angle -\phi - 90° + \tan^{-1}\left(\frac{\omega_o L}{R}\right) = K_0\angle -\theta_0 = a_1^*$$

$$b_2 = \frac{1/2(A\angle-\phi)(-R/L)}{-R/L+j\omega_o} = \frac{1/2(AR/L)}{\sqrt{(R/L)^2+\omega_o^2}}\angle-\phi+\tan^{-1}\left(\frac{\omega_o L}{R}\right) = K_1\angle-\theta_1 = b_1^*$$

$$V_L(s) = \frac{K_o\angle\theta_o}{s-j\omega_o} + \frac{K_o\angle-\theta_o}{s+j\omega_o} + \frac{K_1\angle\theta_1}{s+R/L} + \frac{K_1\angle-\theta_1}{s+R/L} - \frac{v_R(0^-)}{s+R/L}$$

$$V_L(s) = \frac{K_o\angle\theta_o}{s-j\omega_o} + \frac{K_o\angle-\theta_o}{s+j\omega_o} + \frac{2\mathcal{R}e\{K_1\angle\theta_1\}-v_R(0^-)}{s+R/L}$$

$$V_L(s) = \frac{K_o\angle\theta_o}{s-j\omega_o} + \frac{K_o\angle-\theta_o}{s+j\omega_o} + \frac{\dfrac{(AR/L)}{\sqrt{(R/L)^2+\omega_o^2}}\cos\left[\phi-\tan^{-1}\left(\dfrac{\omega_o L}{R}\right)\right]-v_R(0^-)}{s+R/L}$$

$$\Rightarrow v_L(t) = \frac{A\omega_o}{\sqrt{(R/L)^2+\omega_o^2}}\cos\left[\omega_o t+\phi+90°-\tan^{-1}\left(\frac{\omega_o L}{R}\right)\right]u(t)$$

$$+\left[\frac{(AR/L)}{\sqrt{(R/L)^2+\omega_o^2}}\cos\left[\phi-\tan^{-1}\left(\frac{\omega_o L}{R}\right)\right]-v_R(0^-)\right]e^{-\frac{R}{L}t}u(t)$$

39.   Given the circuit shown below:

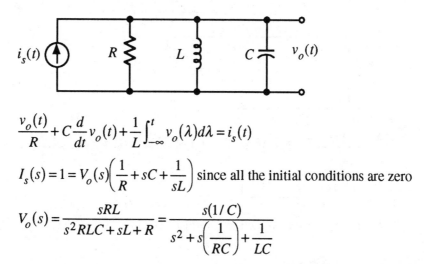

$$\frac{v_o(t)}{R}+C\frac{d}{dt}v_o(t)+\frac{1}{L}\int_{-\infty}^t v_o(\lambda)d\lambda = i_s(t)$$

$$I_s(s) = 1 = V_o(s)\left(\frac{1}{R}+sC+\frac{1}{sL}\right) \text{ since all the initial conditions are zero}$$

$$V_o(s) = \frac{sRL}{s^2RLC+sL+R} = \frac{s(1/C)}{s^2+s\left(\dfrac{1}{RC}\right)+\dfrac{1}{LC}}$$

Since the response is underdamped, there are complex conjugate poles.

$$\Rightarrow s^2+s\left(\frac{1}{RC}\right)+\frac{1}{LC} = \left(s+\frac{1}{2RC}\right)^2+\frac{1}{LC}-\left(\frac{1}{2RC}\right)^2$$

$$V_o(s) = \frac{s/C}{\left(s+\dfrac{1}{2RC}\right)^2+\dfrac{1}{LC}-\left(\dfrac{1}{2RC}\right)^2}$$

Let $\alpha = \dfrac{1}{2RC}$ and $\omega_o^2 = \dfrac{1}{LC}-\left(\dfrac{1}{2RC}\right)^2$

$$V_o(s) = \frac{1}{C}\left[\frac{s+\alpha}{(s+\alpha)^2+\omega_o^2} - \frac{\alpha}{(s+\alpha)^2+\omega_o^2}\right]$$

$$\Rightarrow V_o(s) = \frac{1}{C}\frac{s+\alpha}{(s+\alpha)^2+\omega_o^2} - \frac{\alpha}{\omega_o C}\frac{\omega_o}{(s+\alpha)^2+\omega_o^2}$$

$$\Rightarrow \boxed{v_o(t) = \frac{1}{C}e^{-\alpha t}u(t)\left[\cos(\omega_o t) - \frac{\alpha}{\omega_o}\sin(\omega_o t)\right]}$$

where $\alpha = \dfrac{1}{2RC}$ and $\omega_o^2 = \dfrac{1}{LC} - \left(\dfrac{1}{2RC}\right)^2$

40.   Given the circuit shown below:

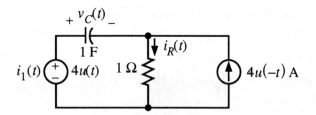

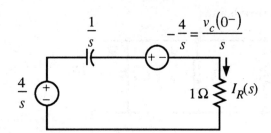

$$\frac{4}{s} = \frac{I_R(s)}{s} - \frac{4}{s} + I_R(s) \Rightarrow \frac{8}{s} = I_R(s)\left(\frac{1+s}{s}\right) \Rightarrow I_R(s) = \frac{8}{s+1}$$

$$\Rightarrow \boxed{i_R(t) = 8e^{-t}u(t)\ \text{A}}$$

41.   Refer to the circuit shown in Exercise 40 with $v_1(t) = 4$ V.

Note that $v_c(0^-) = 0$.

From Exercise 40, then

$$\frac{4}{s} = \frac{I_R(s)}{s} + I_R(s) \Rightarrow I_R(s) = \left(\frac{4}{s+1}\right) \Rightarrow \boxed{i_R(t) = 4e^{-t}u(t)\ \text{A}}$$

42. Refer to the figure shown in Exercise 40 with $v_1(t) = 4u(t-1)$.

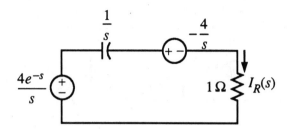

$$\frac{4e^{-s}}{s} = I_R(s)\left(\frac{s+1}{s}\right) - \frac{4}{s} \implies I_R(s) = \frac{4e^{-s}}{s+1} + \frac{4}{s+1}$$

$$\implies \boxed{i_R(t) = 4e^{-(t-1)}u(t-1) + 4e^{-t}u(t)\ \text{A}}$$

43. Refer to the circuit shown in Exercise 40 with $v_1(t) = 8\cos(t+30°)u(t)$.

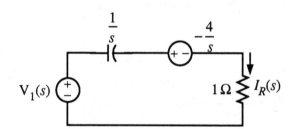

$$V_1(s) = I_R(s)\left(\frac{s+1}{s}\right) - \frac{4}{s} = \frac{4\angle 30°}{s-j} + \frac{4\angle -30°}{s+j}$$

$$\implies I_R(s)\left(\frac{s+1}{s}\right) = \frac{6.928203s - 4}{s^2+1} + \frac{4}{s}$$

$$\implies I_R(s) = \frac{10.928203s^2 - 4s + 4}{(s^2+1)(s+1)} = \frac{a}{s+1} + \frac{bs+c}{s^2+1}$$

$$a = \frac{10.928203(-1)^2 - 4(-1) + 4}{\left[(-1)^2 + 1\right]} = 9.464$$

$$I_R(0) = 4 = 9.464 + c \implies c = -5.464$$

$$\lim_{s\to\infty} sI_R(s) = 10.928 = 9.464 + b$$

$$\implies b = 1.464$$

$$I_R(s) = \frac{10.928203s^2 - 4s + 4}{(s^2+1)(s+1)} = \frac{9.464}{s+1} + \frac{1.464s - 5.464}{s^2+1}$$

$$I_R(s) = \frac{9.464}{s+1} + 1.464\frac{s}{s^2+1} - 5.464\frac{1}{s^2+1}$$

$$\implies \boxed{i_R(t) = \left[9.464e^{-t} + 1.464\cos(t) - 5.464\sin(t)\right]u(t)}$$

44.   Given the circuit shown below:

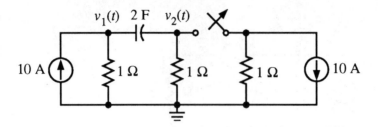

For $t < 0$:

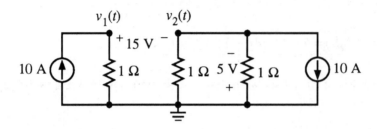

$$v_C(0^-) = 15 \text{ V}$$

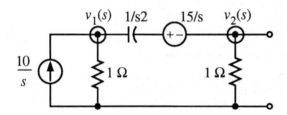

Node 1, KCL: $-\dfrac{10}{s} + V_1(s) + \left[V_1(s) - V_2(s) - 15/s\right]2s = 0$

Node 2, KCL: $\left[V_2(s) + \dfrac{15}{s} - V_1(s)\right]2s + V_2(s) = 0$

Solving these two equations simultaneously yields

$$V_1(s) = \frac{10(5s+1)}{s(4s+1)} = \frac{10}{s} + \frac{5/2}{s+\frac{1}{4}} \Rightarrow \boxed{v_1(t) = 10u(t) + \frac{5}{2}e^{-\frac{1}{4}t}u(t)}$$

$$-\frac{10}{s} + \frac{V_1(s)}{1} + V_2(s) = 0 \Rightarrow V_2(s) = -\frac{5/2}{s+\frac{1}{4}} \Rightarrow \boxed{v_2(t) = -\frac{5}{2}e^{-\frac{1}{4}t}u(t)}$$

45.   (a)   By LT2 Table 14.2, $H(s) = \dfrac{1}{s}$

(b)   We have $V_o(s) = V_s(s)H(s) = \dfrac{1}{s^2}$. Therefore, by LT3 of Table 14.2, $v_o(t) = tu(t)$

46.   (a)   From Example 41.1 or LT5 of Table 14.2, we have $H(s) = \dfrac{1}{s+3}$

(b)   Except for scaling, the steps are indentical to those in Example 14.20.

The result is $v_o(t) = \frac{1}{3}\left[1 - e^{-3t}\right]u(t)$

47.   Convolution is a linear operation.

Therefore, from Exercise 46 we have $v_o(t) = e^{-3t}u(t) + \frac{1}{3}\left[1 - e^{-3t}\right]u(t)$

CHAPTER 15

Fourier Series

Exercises

1. (a) $v_5(t) = \sum_{n=1,3,5,...}^{5} \dfrac{8}{(\pi n)^2} \sin(2\pi n f_1 t)$

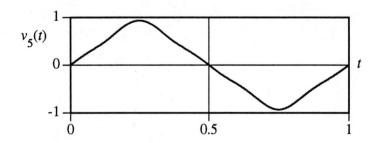

(b) Given the circuit shown below;

$$V_L = \frac{j2\pi n f_1 L}{1 + j2\pi n f_1 L} \cdot V_s$$

For $f = f_1 = 1\,\text{Hz}$: $V_{L1} = \dfrac{j4\pi}{1 + j4\pi}\left(\dfrac{8}{\pi^2}\right) = 0.808\angle 4.55°$

$\Rightarrow v_{L1}(t) = 0.808\sin(2\pi t + 4.55°)\,\text{V}$

For $f = 3f_1 = 3\,\text{Hz}$: $V_{L2} = \dfrac{j12\pi}{1 + j12\pi}\left(\dfrac{8}{9\pi^2}\right) = 0.090\angle 1.52°$

$\Rightarrow v_{L2}(t) = 0.090\sin(6\pi t + 1.52°)\,\text{V}$

For $f = 5f_1 = 5\,\text{Hz}$: $\mathbf{V}_{L3} = \dfrac{j20\pi}{1+j20\pi}\left(\dfrac{8}{25\pi^2}\right) = 0.033\angle 0.91°$

$\Rightarrow v_{L3}(t) = 0.090\sin(6\pi t + 1.52°)\,\text{V}$

$v_L(t) = v_{L1}(t) + v_{L2}(2) + v_{L3}(t)$

$v_L(t) = 0.808\sin(2\pi t + 4.55°) + 0.09\sin(6\pi t + 1.52°) + 0.09\sin(6\pi t + 1.52°)\,\text{V}$

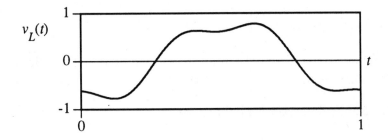

2.　　$v_7(t) = \displaystyle\sum_{n=1,3,5,\dots}^{7} \dfrac{8}{(\pi n)^2}\sin(2\pi n f_1 t)$

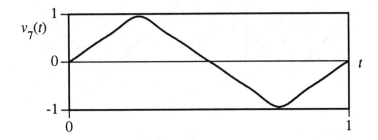

3.　　$v(t) = 3 + 5\cos(2\pi 10^3 t) + 7\sin(6\pi 10^3 t)$

$\Rightarrow$ $\boxed{T = 1\,\text{ms},\ a_0 = 3,\ a_1 = 5,\ b_s = 7,\ \text{and all other coefficients are } 0}$

4.　　Right-hand side of Eq. (15.9) $\Rightarrow 3^2 + \dfrac{1}{2}(5^2 + 7^2) = 46$

5.　　(a)

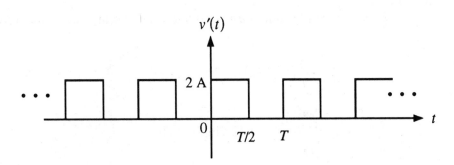

　　(b)　　No, $v'(t)$ does not have odd symmetry.

　　(c)　　No, $v'(t)$ does not display half-wave symmetry.

(d) $v'(t) = \text{A} + \sum_{n=1}^{\infty} \dfrac{4\text{A}}{n\pi}\sin\left(2\pi f_1 t\right)$

(e)   Only the dc component is different, since the average value is the only thing that has been changed.

6.   (a) $v_1(t)$ and $v_2(t)$ have no sine components

       (b) $v_3(t)$ and $v_4(t)$ hae no cosine components

       (c) $v_2(t)$ and $v_4(t)$ have only frequencies that are odd multiples of f_1, since they display half-wave symmetry.

7.   Given the signal shown below:

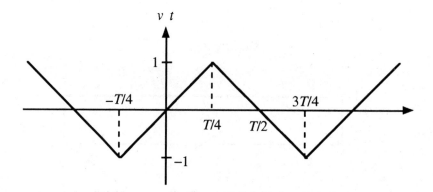

$$v(t) = \begin{cases} -\dfrac{4}{T}\left(t + \dfrac{T}{2}\right) & -\dfrac{T}{2} \le t < -\dfrac{T}{4} \\[2mm] \dfrac{4}{T}t & -\dfrac{T}{4} \le t < \dfrac{T}{4} \\[2mm] -\dfrac{4}{T}\left(t - \dfrac{T}{2}\right) & \dfrac{T}{4} \le t < \dfrac{T}{2} \end{cases}$$

$a_0 = 0$ since the average value is zero.

$a_n = 0$ since the function is odd and thus contains only sine terms.

$$b_n = \frac{2}{T}\int_{-T/2}^{-T/4} -\frac{4}{T}(t + T/2)\sin\left(2\pi n f_1 t\right)dt + \frac{2}{T}\int_{-T/4}^{T/4} \frac{4}{T}t\sin\left(2\pi n f_1 t\right)dt$$

$$+ \frac{2}{T}\int_{-T/2}^{-T/4} -\frac{4}{T}(t - T/2)\sin\left(2\pi n f_1 t\right)dt$$

With the help of the identity $\displaystyle\int x\sin(ax)dx = \dfrac{\sin(ax) - ax\cos(ax)}{a^2}$

$$b_n = (-1)^{\frac{n-1}{2}} \frac{8A}{(n\pi)^2}, \text{ for } n \text{ odd and } b_0 = 0$$

$$\Rightarrow \quad v(t) = \sum_{n=1}^{\infty} (-1)^{\frac{n-1}{2}} \frac{8A}{(n\pi)^2} \sin(2\pi n f_1 t), \text{ for } n \text{ odd}$$

8.   Given the signal shown below:

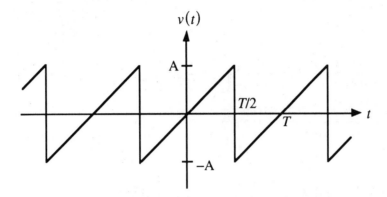

$$v(t) = \frac{2A}{T} t \text{ for } -\frac{T}{2} \le t \le \frac{T}{2}$$

$a_0 = 0$ since the average value is zero.

$a_n = 0$ since the function is odd.

$$b_n = \frac{2}{T} \int_{-\frac{T}{2}}^{\frac{T}{2}} \frac{2A}{T} t \sin(2\pi n f_1 t) dt = \frac{4A}{T^2} \int_{-\frac{T}{2}}^{\frac{T}{2}} t \sin(2\pi n f_1 t) dt$$

With the help of the identity

$$\int x \sin(ax) dx = \frac{\sin(ax) - ax \cos(ax)}{a^2},$$

$$b_n = \frac{4A}{T^2} \left[\frac{\sin(2\pi n f_1 t) - 2\pi n f_1 t \cos(2\pi n f_1 t)}{(2\pi n f_1)^2} \right]_{-\frac{T}{2}}^{\frac{T}{2}} = -\frac{2n\pi A}{(n\pi)^2} \cos(n\pi)$$

$$\Rightarrow \quad b_n = \frac{2A}{n\pi}, \text{ for } n \text{ odd} \quad : \quad b_n = -\frac{2A}{n\pi}, \text{ for } n \text{ even} \quad \text{or} \quad b_n = (-1)^{n+1} \frac{2A}{n\pi}$$

$$\Rightarrow \quad v(t) = \sum_{n=1}^{\infty} (-1)^{n+1} \frac{2A}{n\pi} \sin(2\pi n f_1 t)$$

9.   For the waveform shown in Fig. 15.1,

147

(a) $A_n = \sqrt{a_n^2 + b_n^2} = \sqrt{0 + \left(\dfrac{8}{n\pi}\right)^2} \Rightarrow \boxed{A_n = \dfrac{8}{n\pi}, \text{ for } n \text{ odd}}$

$$\boxed{\theta = \tan^{-1}\left(-\dfrac{b_n}{a_n}\right) = (-1)^{\frac{n+1}{2}} \cdot 90°, \text{ for } n \text{ odd}}$$

(b) $P = \dfrac{1}{T}\displaystyle\int_{-\frac{T}{2}}^{-\frac{T}{4}}\left(-\dfrac{4}{T}\right)^2\left(t + \dfrac{T}{2}\right)^2 dt + \dfrac{1}{T}\int_{-\frac{T}{4}}^{\frac{T}{4}}\left(\dfrac{4}{T}t\right)^2 dt + \dfrac{1}{T}\int_{\frac{T}{4}}^{\frac{T}{2}}\left(-\dfrac{4}{T}\right)^2\left(t - \dfrac{T}{2}\right)^2 dt$

$P = \dfrac{16}{T^3}\left[\dfrac{1}{3}t^3 + \dfrac{T}{2}t^2 + \dfrac{T^2}{4}t^3\right]_{-\frac{T}{2}}^{-\frac{T}{4}} + \dfrac{16}{3T^3}t^3\Big|_{-\frac{T}{4}}^{\frac{T}{4}} + \dfrac{16}{T^3}\left[\dfrac{1}{3}t^3 - \dfrac{T}{2}t^2 + \dfrac{T^2}{4}t^3\right]_{\frac{T}{4}}^{\frac{T}{2}}$

$\Rightarrow \boxed{P = \dfrac{1}{3}\text{W}}$

(c) $P_4 = \dfrac{1}{2}A_1^2 + \dfrac{1}{2}A_3^2 = \dfrac{1}{2}\left(\dfrac{64}{\pi^4}\right) + \dfrac{1}{2}\left(\dfrac{64}{81\pi^2}\right) = 0.3326$

This is 99.77 percent of the total power.

10.   Given the signal shown below:

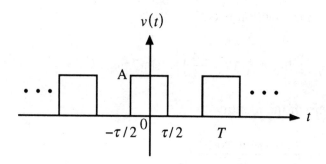

(a) $V_n = \dfrac{1}{T}\displaystyle\int_{-\frac{\tau}{2}}^{\frac{\tau}{2}} A e^{j2\pi n f_1 t}\,dt = \dfrac{A}{j2\pi n f_1 T}\left(e^{j\pi n f_1 \tau} - e^{-j\pi n f_1 \tau}\right)$

$\Rightarrow \boxed{V_n = \dfrac{A}{n\pi f_1 T}\sin(n\pi f_1 \tau),\ n \neq 0} \ :\ V_o = \dfrac{1}{T}\displaystyle\int_{-\frac{\tau}{2}}^{\frac{\tau}{2}} A\,dt \Rightarrow \boxed{V_o = \dfrac{A\tau}{T}}$

or, multiplying the top and bottom of the equation for V_n by τ yields

$$\boxed{V_n = \dfrac{A\tau}{T}\dfrac{\sin(n\pi f_1 \tau)}{n\pi f_1 \tau}, \text{ for } n \neq 0}$$

(b) $a_n = 2\mathcal{Re}\{V_n\} = \dfrac{2A\tau}{T}\dfrac{\sin(n\pi f_1 \tau)}{n\pi f_1 \tau} \ :\ b_n = 2\mathcal{Im}\{V_n\} = 0$

$a_0 = V_0 = \dfrac{A\tau}{T}$ These are the same results obtained in Example 15.3.

11. $V_n = \dfrac{1}{T}\int_T v(t)e^{-j2\pi n f_1 t}\,dt = \dfrac{1}{T}\int_T v(t)\cos(2\pi n f_1 t)\,dt - j\dfrac{1}{T}\int_T v(t)\sin(2\pi n f_1 t)\,dt$

(a) If $v(t)$ is an even function then $v(t)\sin(2\pi n f_1 t)$ is an odd function of time, and therefore the second integral will be zero. Therefore V_n is real when $v(t)$ is real and even.

(b) If $v(t)$ is a real, odd function of time, then the first integral will be zero. Therefore V_n is imaginary.

12. $v(t) = \displaystyle\sum_{i=-\infty}^{\infty}\delta(t - iT)$

$$\boxed{V_n = \frac{1}{T}\int_{0^-}^{0^+}\delta(t)e^{j2\pi n f_1 t}\,dt = \frac{1}{T}e^0 = \frac{1}{T} \quad \text{for } n = 0, \pm1, \pm2, \pm3, \ldots}$$

13. $v(t) = \displaystyle\sum_{n=1,3,5,\ldots}^{\infty}(-1)^{(n-1)/2}\frac{8}{(\pi n)^2}\sin(2\pi n f_1 t)$

$= \displaystyle\sum_{n=1,3,5,\ldots}^{\infty}(-1)^{(n-1)/2}\frac{8}{(\pi n)^2}\left(\frac{e^{j2\pi n f_1 t} - e^{-j2\pi n f_1 t}}{j2}\right)$

$= \displaystyle\sum_{n=1,3,5,\ldots}^{\infty}(-1)^{(n-1)/2}\frac{8}{j2(\pi n)^2}e^{j2\pi n f_1 t} - \sum_{n=1,3,5,\ldots}^{\infty}(-1)^{(n-1)/2}\frac{8}{j2(\pi n)^2}e^{-j2\pi n f_1 t}$

$= \displaystyle\sum_{n=1,3,5,\ldots}^{\infty}(-1)^{(n-1)/2}\frac{4}{j(\pi n)^2}e^{j2\pi n f_1 t} + \sum_{n=1,3,5,\ldots}^{\infty}(-1)^{(n-1)/2}\frac{4}{j(\pi n)^2}e^{-j2\pi n f_1 t}$

$\Rightarrow \boxed{V_n = j\dfrac{4}{(n\pi)^2}(-1)^{(n+1)/2}, \text{ for } n \text{ odd}} \quad \Rightarrow \boxed{V_n = 0, \text{ for } n \text{ even}}$

14. (a)

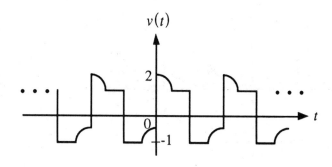

(b) $\boxed{V_n = -j\dfrac{2A}{n\pi}, \text{ for } n \text{ odd}, n \neq \pm 1 \; : \; V_n = \dfrac{A}{4} - j\dfrac{2A}{n\pi}, \text{ for } n = \pm 1}$

$\boxed{V_n = \dfrac{(A/\pi)(-1)^{(n-1)/2}}{1-n^2}, \text{ for } n \text{ even}}$

$\Rightarrow V_0 = \dfrac{A}{\pi}, V_2 = \dfrac{A}{3\pi} = V_{-2}, V_4 = V_{-4} = -\dfrac{A}{15\pi}$

$V_1 = \dfrac{A}{4} - j\dfrac{2A}{\pi} \; : \; V_{-1} = \dfrac{A}{4} + j\dfrac{2A}{\pi} \; : \; V_3 = -j\dfrac{2A}{3\pi} \; : \; V_{-3} = j\dfrac{2A}{3\pi}$

(c)

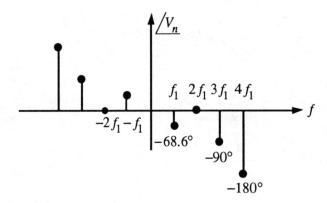

15. $g(t) = Ae^{t/\tau}$

(a)

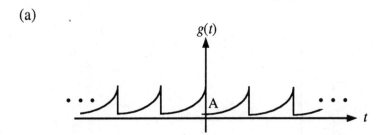

(b) From FS3, $V_n = \dfrac{A\tau}{T} \cdot \dfrac{1-e^{-T/\tau}}{1+j2\pi n\tau/T}$

From property P2, $V_{-n} = G_n = \dfrac{A\tau}{T} \cdot \dfrac{1 - e^{-T/\tau}}{1 - j2\pi n\tau/T}$

$$G_n = \frac{A\tau}{T} \cdot \frac{1 - e^{-T/\tau}}{\sqrt{1 + \left(4\pi^2 n^2 \tau^2 / T^2\right)}} \angle \tan^{-1}\left(\frac{2\pi n\tau}{T}\right)$$

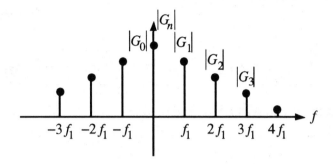

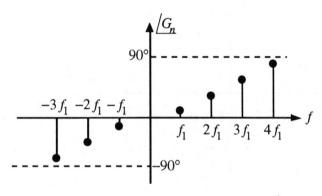

16. $\qquad v_d(t) = v\left(t - \dfrac{T}{2}\right)$

(a)

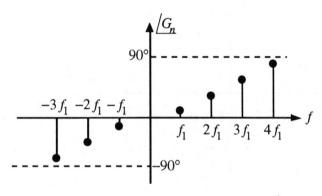

(b) $\qquad V_{dn} = \dfrac{(A/\pi)\cos(n\pi/2)}{1 - n^2} e^{-j\frac{2\pi n}{T}\frac{T}{2}} = V_n e^{-j\pi n}, n \neq \pm 1$

$$\boxed{V_{dn} = \frac{(A/\pi)\cos(n\pi/2)}{1 - n^2} e^{-j\pi n}, n \neq \pm 1 \; : \; V_{dn} = \frac{A}{4} e^{-j\pi n} \; n = \pm 1}$$

(c) The full-rectified cosine wave is simply the sum of a half-rectified cosine wave and a half-rectified cosine wave shifted by $T/2$.

Chapter 15

$$\Rightarrow V_{FR_n} = V_n + V_{dn} = \frac{(A/\pi)\cos(n\pi/2)}{(1-n)^2}\left(1+e^{-j\pi n}\right), n \neq \pm 1$$

Since $1+e^{-j\pi n}=0$ for the same n as $\cos(n\pi/2)$, this can be written as

$$V_{FR_n} = \frac{(2A/\pi)\cos(n\pi/2)}{(1-n)^2}, n \neq \pm 1$$

For $n = \pm 1$: $\quad V_{FR_n} = V_n + V_{dn} = \frac{A}{4}\left(1+e^{-j\pi n}\right) = 0$ for $n = \pm 1$

This result is the same as FS09 in the table.

(d)

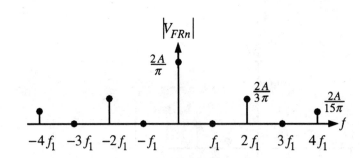

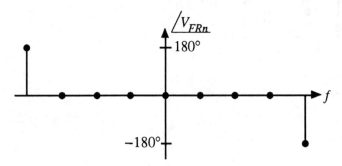

17.   The triangular waveform from Figure 15.1 is shown below.

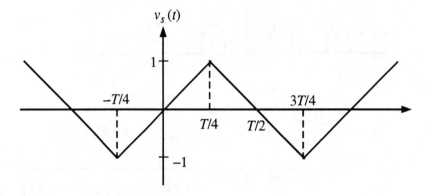

$$V_n = j\frac{4}{(\pi n)^2}(-1)^{\frac{n+1}{2}}, \text{ for } n \text{ odd}$$

The derivative of this signal is shown below.

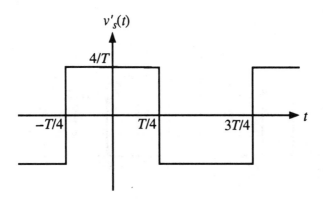

$$\frac{d}{dt}v(t) \Leftrightarrow j2\pi n f_1 V_n \Rightarrow \boxed{V_{sn} = -\frac{8\pi n f_1}{(n\pi)^2}(-1)^{\frac{n+1}{2}}, \text{ for } n \text{ odd}}$$

This disagrees with FSO1 because the square wave in FS01 has odd symmetry and this has even symmetry

18.   For series RC circuit with the output across the capacitor and the values given,

$$H(n) = \frac{1}{1 + j0.1n\pi} = \frac{1}{\sqrt{1 + (0.1n\pi)^2}} \angle \tan^{-1}(0.1n\pi)$$

The input has the form $V_n = \dfrac{\sin(0.1n\pi)}{0.1n\pi}$.

The output then is the product of the two.

$$\boxed{V_{on} = \frac{1}{\sqrt{1 + (0.1n\pi)^2}} \frac{\sin(0.1n\pi)}{0.1n\pi} \angle -\tan^{-1}(0.1n\pi)}$$

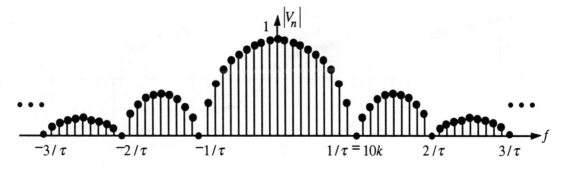

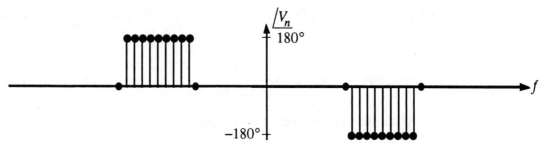

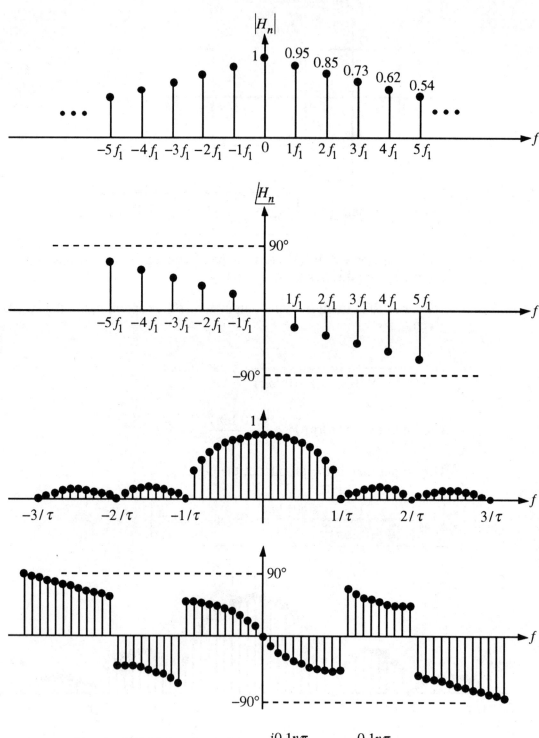

19.   For the RC high-pass, $H(n) = \dfrac{j0.1n\pi}{1+j0.1n\pi} = \dfrac{0.1n\pi}{\sqrt{1+(0.1n\pi)^2}} \angle 90° - \tan^{-1}(0.1n\pi)$

The input has the form $V_n = \dfrac{\sin(0.1n\pi)}{0.1n\pi}$.

The output then is the product of the two.

$$V_{on} = \frac{\sin(0.1n\pi)}{0.1n\pi\sqrt{1+(0.1n\pi)^2}} \angle 90° - \tan^{-1}(0.1n\pi)$$

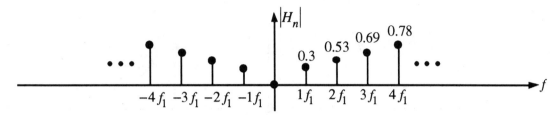

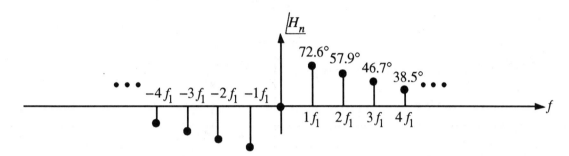

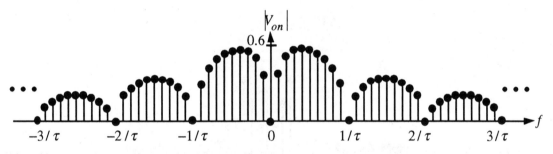

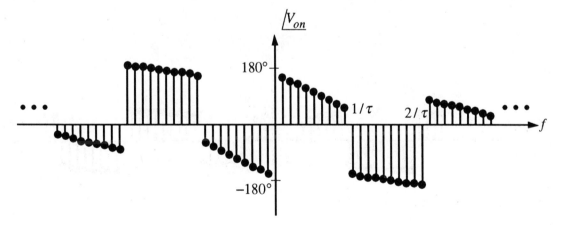

20. Refer to the circuit shown in Fig. 15.15.

$$H(nf_1) = \frac{1}{1 + jn10^{-3}f_1 + \frac{9\times10^3}{jnf_1}}$$

$$T = 1 \text{ ms}, f_1 = 1000 \text{ Hz} \Rightarrow H(nf_1) = \frac{1}{1 + jn - j\frac{9}{n}} = \frac{1}{1 + j\left(n - \frac{9}{n}\right)}$$

$$\Rightarrow \boxed{V_{on} = \frac{1}{\sqrt{1+\left(n-\frac{9}{n}\right)^2}} \frac{\sin(0.1n\pi)}{0.1n\pi} \angle -\tan^{-1}\left(n-\frac{9}{n}\right)}$$

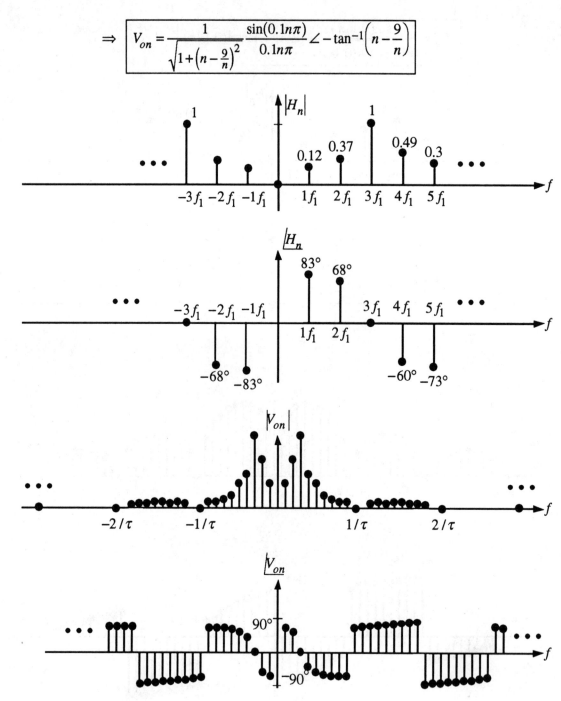

21.   Given the circuit below.

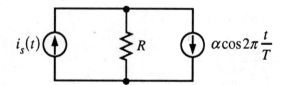

(a) $P = \frac{1}{T}\int_T \left|i_s(t) - \alpha\cos\left(2\pi f_1 t\right)\right|^2 Rdt$ where $f_1 = \frac{1}{T}$

$$P = \frac{R}{T}\int_T i_s^2(t)dt - 2\alpha\frac{R}{T}\int_T i_s(t)\cos\left(2\pi f_1 t\right)dt + \alpha^2 R\frac{1}{T}\frac{T}{2}$$

$$P = \frac{R}{T}\int_T i_s^2(t)dt - 2\alpha\frac{R}{T}\int_T i_s\cos\left(2\pi f_1 t\right)dt + \frac{\alpha^2 R}{2}$$

$$\frac{d}{d\alpha}P = -\frac{2R}{T}\int_T i_s(t)\cos\left(2\pi f_1 t\right)dt + \alpha R = 0 \Rightarrow \boxed{\alpha = \frac{2}{T}\int_T i_s(t)\cos\left(2\pi f_1 t\right)dt = a_1}$$

(b) $\quad P = \frac{R}{T}\int_T i_s^2(t)dt - 2\alpha\frac{R}{T}\int_T i_s(t)\cos\left(2\pi f_1 t\right)dt + \frac{\alpha^2 R}{2}$

$$P = \frac{R}{T}\int_T i_s^2(t)dt - \frac{4R}{T^2}\int_T i_s\cos\left(2\pi f_1 t\right)dt \int_T i_s(t)\cos\left(2\pi f_1 t\right)dt + \frac{\alpha^2 R}{2}$$

$$P = \frac{R}{T}\int_T i_s^2(t)dt - \alpha^2 R + \frac{\alpha^2 R}{2} \Rightarrow \boxed{P = \frac{R}{T}\int_T i_s^2(t)dt - \frac{\alpha^2 R}{2}}$$

22. e^{-3t}, $0 \le t \le T$ satisfies all three Dirichlet conditions.

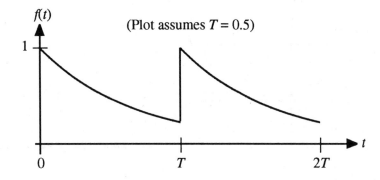

23. $1/t$, $0 < t < T$ satisfies all three Dirichlet conditions.

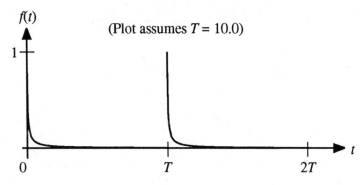

24. $\cos(1/t)$, $0 \le t < T$ violates condition 2.

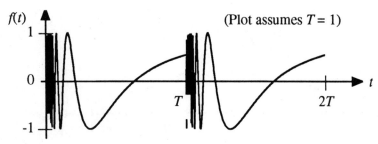

25. $u(t)+u(t-0.5T)+u(t-0.75T)+u(t-0.625T)+\dots,\quad 0\le t<T$ violates
conditon 3.

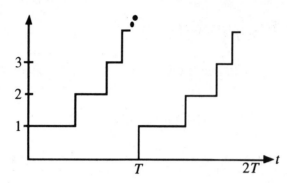

26. $v(t)=\delta(t)\ \Rightarrow\ \mathbf{V}(f)=\int_{-\infty}^{\infty}\delta(t)e^{-j2\pi ft}dt=1$

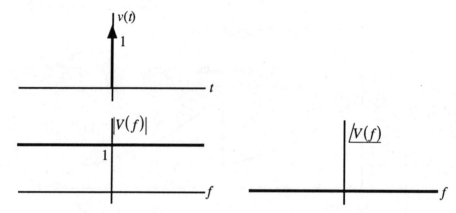

27. $v(t)=e^{bt}u(-t),\ b>0$

$$\mathbf{V}(f)=\int_{-\infty}^{0}e^{bt}e^{-j2\pi ft}dt=\int_{-\infty}^{0}e^{(b-j2\pi f)t}dt=\frac{1}{b-j2\pi f}$$

(Plot assumes $b=1$)

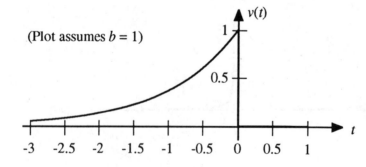

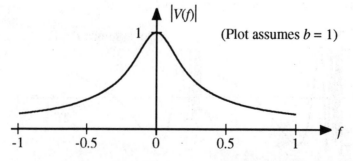

(Plot assumes $b = 1$)

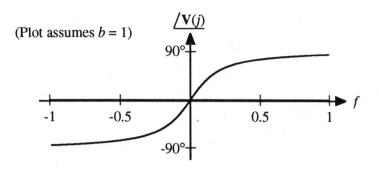

28. $v(t) = e^{at}u(t) + e^{bt}u(-t), \quad a < 0, \ b > 0$

$$\Rightarrow \mathbf{V}(f) = \frac{1}{j2\pi f - a} + \frac{1}{b - j2\pi f} = \frac{b - a}{-ab + (2\pi f)^2 + j2\pi f(b + a)}$$

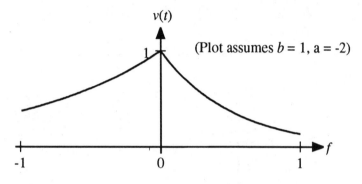

(Plot assumes $b = 1$, a = -2)

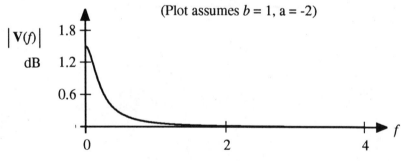

(Plot assumes $b = 1$, a = -2)

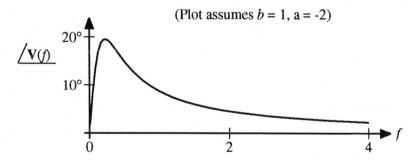

(Plot assumes $b = 1$, a = -2)

29. $\mathbf{V}(f) = \delta(f) \ \Rightarrow \ v(t) = 1$

30. $\mathbf{V}(f) = \frac{1}{2}A\delta(f - f_0) + \frac{1}{2}A\delta(f + f_0) \ \Rightarrow \ v(t) = A\cos(2\pi f_o t)$

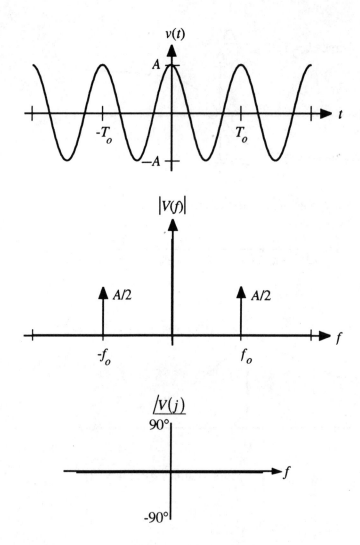

CHAPTER 16

Equivalent Circuits for Three-Terminal Networks and Two-Port Networks

Exercises

1. Since there are no independent sources, $\mathbf{V}_{1oc} = \mathbf{V}_{2oc} = 0$.

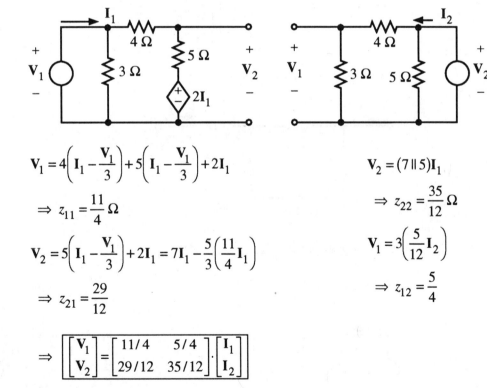

Find z_{11} and z_{21}:

Find z_{12} and z_{22}:

$$V_1 = 4\left(I_1 - \frac{V_1}{3}\right) + 5\left(I_1 - \frac{V_1}{3}\right) + 2I_1$$

$$\Rightarrow z_{11} = \frac{11}{4}\,\Omega$$

$$V_2 = 5\left(I_1 - \frac{V_1}{3}\right) + 2I_1 = 7I_1 - \frac{5}{3}\left(\frac{11}{4}I_1\right)$$

$$\Rightarrow z_{21} = \frac{29}{12}$$

$$V_2 = (7\,\|\,5)I_1$$

$$\Rightarrow z_{22} = \frac{35}{12}\,\Omega$$

$$V_1 = 3\left(\frac{5}{12}I_2\right)$$

$$\Rightarrow z_{12} = \frac{5}{4}$$

$$\Rightarrow \begin{bmatrix} V_1 \\ V_2 \end{bmatrix} = \begin{bmatrix} 11/4 & 5/4 \\ 29/12 & 35/12 \end{bmatrix} \cdot \begin{bmatrix} I_1 \\ I_2 \end{bmatrix}$$

2. Given the circuit shown below:

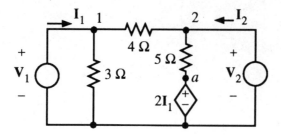

Node 1, KCL: $-I_1 + \dfrac{V_1}{3} + \dfrac{V_1 - V_2}{4} = 0 \;\Rightarrow\; V_2 = \dfrac{7}{3}V_1 - 4I_1$ (1)

Node 2, KCL: $-V_1\left(-\dfrac{29}{60}\right) + V_2\left(\dfrac{11}{20}\right) = I_2 \;\Rightarrow\; V_1 = \dfrac{33}{29}V_2 - \dfrac{60}{29}I_2$ (2)

Substituting Eq. (1) into Eq. (2) yields the following:

$$V_1 = \frac{11}{4}I_1 + \frac{5}{4}I_2$$

Substituting Eq. (2) into Eq. (1) yields the following:

$$V_2 = \frac{29}{12}I_2 + \frac{35}{12}I_2$$

$$\Rightarrow \boxed{\begin{bmatrix} V_1 \\ V_2 \end{bmatrix} = \begin{bmatrix} 11/4 & 5/4 \\ 29/12 & 35/12 \end{bmatrix} \cdot \begin{bmatrix} I_1 \\ I_2 \end{bmatrix}}$$

3.　　　　Given the circuit shown below:

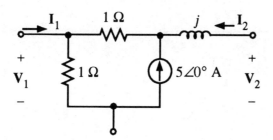

$\mathbf{V}_{1oc} = 5\angle 0°$ and $\mathbf{V}_{2oc} = 5 + 5 = 10\angle 0°$ V

Find z_{11} and z_{21}:　　　　　　　　　　　　　　　Find z_{12} and z_{22}:

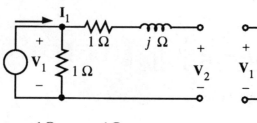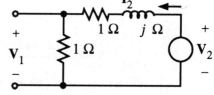

$z_{11} = 1\,\Omega : z_{21} = 1\,\Omega$　　　　　　　　　　$z_{12} = 1\,\Omega : z_{22} = 2 + j\,\Omega$

$$\Rightarrow \begin{bmatrix} \mathbf{V}_1 \\ \mathbf{V}_2 \end{bmatrix} = \begin{bmatrix} 1 & 1 \\ 1 & 2+j \end{bmatrix} \cdot \begin{bmatrix} \mathbf{I}_1 \\ \mathbf{I}_2 \end{bmatrix} + \begin{bmatrix} 5\angle 0^\circ \\ 10\angle 0^\circ \end{bmatrix}$$

4.   Given the circuit shown below:

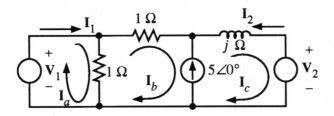

Mesh 1, KVL: $\mathbf{V}_1 = 1\left(\mathbf{I}_a - \mathbf{I}_b\right)$ (1)

Mesh 2-3, KCL: $\mathbf{I}_c - \mathbf{I}_b = 5 \;\Rightarrow\; \mathbf{I}_b = \mathbf{I}_c - 5$ (2)

Mesh 2-3, KVL: $\mathbf{I}_b + j\mathbf{I}_c + \mathbf{V}_2 + 1\left(\mathbf{I}_b - \mathbf{I}_a\right) = 0$ (3)

Substituting Eq. (2) into Eq. (1) yields $\mathbf{V}_1 = \mathbf{I}_1 + \mathbf{I}_2 + 5\angle 0^\circ$

Substituting Eq. (2) into Eq. (3) yields $\mathbf{V}_2 = \mathbf{I}_1 + (2+j)\mathbf{I}_2 + 10\angle 0^\circ$

$$\Rightarrow \begin{bmatrix} \mathbf{V}_1 \\ \mathbf{V}_2 \end{bmatrix} = \begin{bmatrix} 1 & 1 \\ 1 & 2+j \end{bmatrix} \cdot \begin{bmatrix} \mathbf{I}_1 \\ \mathbf{I}_2 \end{bmatrix} + \begin{bmatrix} 5\angle 0^\circ \\ 10\angle 0^\circ \end{bmatrix}$$

5.   Since there are no independent sources, the short circuit currents are zero.

Find y_{11} and y_{21}:   Find y_{12} and y_{22}:

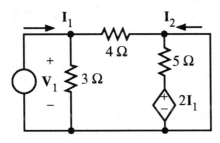 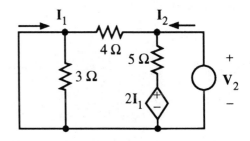

$\mathbf{V}_1 = (3 \| 4)\mathbf{I}_1$ $\mathbf{I}_1 = -\dfrac{\mathbf{V}_2}{4}$

      $\Rightarrow y_{11} = \dfrac{7}{12}S$ $\Rightarrow y_{12} = -\dfrac{1}{4}$

$$\text{KCL:} \quad I_1 - \frac{V_1}{3} + \frac{2}{5}I_1 + I_2 = 0 \qquad\qquad I_5 = \frac{V_2 - 2I_1}{5} = \frac{V_2 - 2(-V_2/4)}{5}$$

$$\Rightarrow I_2 = \left[-\frac{7}{12} + \frac{1}{3} - \frac{2}{5}\left(\frac{7}{12}\right) \right] V_1 \qquad I_5 = \frac{3}{10}V_2 \; : \; I_2 = \frac{3}{10}V_2 + \frac{V_2}{4} = \frac{11}{20}V_2$$

$$\Rightarrow y_{21} = -\frac{29}{60} \qquad\qquad\qquad\qquad \Rightarrow y_{22} = \frac{11}{20}$$

$$\Rightarrow \begin{bmatrix} I_1 \\ I_2 \end{bmatrix} = \begin{bmatrix} \frac{7}{12} & -\frac{1}{4} \\ -\frac{29}{60} & \frac{11}{20} \end{bmatrix} \cdot \begin{bmatrix} V_1 \\ V_2 \end{bmatrix}$$

6. Given the circuit shown below:

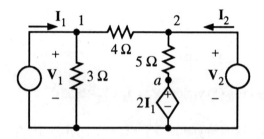

Node 1, KCL: $\quad -I_1 + \dfrac{V_1}{3} + \dfrac{V_1 - V_2}{4} = 0 \;\Rightarrow\; I_1 = \dfrac{7}{12}V_1 - \dfrac{1}{4}V_2$

Node 2, KCL: $\quad \dfrac{V_2 - V_1}{4} + \dfrac{V_2 - 2I_1}{5} - I_2 = 0$

Substituting the equation for I_1 from above yields

$$I_2 = -\frac{29}{60}V_1 + \frac{11}{20}V_2$$

$$\Rightarrow \begin{bmatrix} I_1 \\ I_2 \end{bmatrix} = \begin{bmatrix} \frac{7}{12} & -\frac{1}{4} \\ -\frac{29}{60} & \frac{11}{20} \end{bmatrix} \cdot \begin{bmatrix} V_1 \\ V_2 \end{bmatrix}$$

7. Given the circuit shown below:

$$I_{1sc} = \frac{j}{1+j}(5) = \frac{5}{\sqrt{2}} \angle 45° \text{ A} \; : \; I_{2sc} = \frac{1}{1+j}(5) = \frac{5}{\sqrt{2}} \angle -45° \text{ A}$$

Find y_{11} and y_{21}:

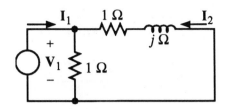

Find y_{12} and y_{22}:

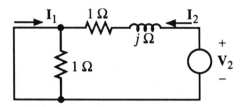

$$\mathbf{V}_1 = \left[1 \| (1+j) \right] \mathbf{I}_1$$

$$\Rightarrow y_{11} = \frac{3}{2} - j\frac{1}{2}$$

$$\mathbf{V}_1 = -\mathbf{I}_2(1+j)$$

$$\Rightarrow y_{21} = -\frac{1}{2} + j\frac{1}{2}$$

$$\mathbf{V}_2 = (1+j)\mathbf{I}_1$$

$$\Rightarrow y_{22} = \frac{1}{2} - j\frac{1}{2}$$

$$\mathbf{V}_2 = -\mathbf{I}_1(1+j)$$

$$\Rightarrow y_{12} = -\frac{1}{2} + j\frac{1}{2}$$

$$\Rightarrow \begin{bmatrix} \mathbf{I}_1 \\ \mathbf{I}_2 \end{bmatrix} = \begin{bmatrix} \frac{3}{2} - j\frac{1}{2} & -\frac{1}{2} + j\frac{1}{2} \\ -\frac{1}{2} + j\frac{1}{2} & \frac{1}{2} - j\frac{1}{2} \end{bmatrix} \cdot \begin{bmatrix} \mathbf{V}_1 \\ \mathbf{V}_2 \end{bmatrix} + \begin{bmatrix} 3.54\angle 45° \\ 3.54\angle -45° \end{bmatrix}$$

8. Given the circuit shown below:

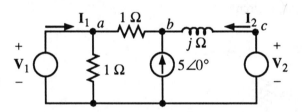

Node a , KCL: $-\mathbf{I}_1 + 2\mathbf{V}_1 = \mathbf{V}_b$ (1)

Node b, KCL: $\dfrac{\mathbf{V}_b - \mathbf{V}_1}{1} - 5 + \dfrac{\mathbf{V}_b - \mathbf{V}_2}{j} = 0$ (2)

Substituting Eq. (1) into Eq. (2) yields the following:

$$\mathbf{I}_1 = \left(\frac{3}{2} - j\frac{1}{2} \right)\mathbf{V}_1 + \left(-\frac{1}{2} + j\frac{1}{2} \right)\mathbf{V}_2 - \frac{5}{\sqrt{2}}\angle 45°$$

Node c, KCL: $\mathbf{I}_2 = -j\mathbf{V}_2 + j\mathbf{V}_b = j2\mathbf{V}_1 - j\mathbf{I}_1 - j\mathbf{V}_2$

Substituting the equation for $\mathbf{I}_1$ from above yields

$$I_2 = \left(-\frac{1}{2} + j\frac{1}{2}\right)V_1 + \left(\frac{1}{2} - j\frac{1}{2}\right)V_2 - \frac{5}{\sqrt{2}} \angle -45°$$

$$\Rightarrow \begin{bmatrix} I_1 \\ I_2 \end{bmatrix} = \begin{bmatrix} \frac{3}{2} - j\frac{1}{2} & -\frac{1}{2} + j\frac{1}{2} \\ -\frac{1}{2} + j\frac{1}{2} & \frac{1}{2} - j\frac{1}{2} \end{bmatrix} \cdot \begin{bmatrix} V_1 \\ V_2 \end{bmatrix} + \begin{bmatrix} 3.54\angle 45° \\ 3.54\angle -45° \end{bmatrix}$$

9.   Given the following impedance matrix:

$$Z = \begin{bmatrix} 9/4 & 5/4 \\ 5/4 & 35/12 \end{bmatrix}$$

then $|Z| = \begin{vmatrix} 9/4 & 5/4 \\ 5/4 & 35/12 \end{vmatrix} = 5$

$$Y = \begin{bmatrix} 35/60 & -5/20 \\ -5/20 & 9/20 \end{bmatrix} = \begin{bmatrix} 7/12 & -1/4 \\ -1/4 & 9/20 \end{bmatrix}$$

$$I_{sc} = \begin{bmatrix} 7/12 & -1/4 \\ -1/4 & 9/20 \end{bmatrix} \cdot \begin{bmatrix} 3 \\ 7 \end{bmatrix} = \begin{bmatrix} 0 \\ 12/5 \end{bmatrix}$$

Yes, the results agree.

10.   The circuit shown below has both a z-parameter equivalent circuit and a y-parameter equivalent circuit.

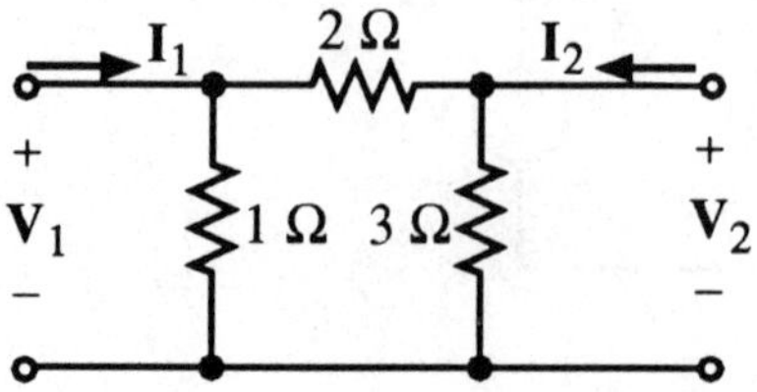

$$Z = \begin{bmatrix} z_{11} & z_{12} \\ z_{21} & z_{22} \end{bmatrix} = \begin{bmatrix} 5/6 & 1/2 \\ 1/2 & 3/2 \end{bmatrix} \quad : \quad Y = \begin{bmatrix} y_{11} & y_{12} \\ y_{21} & y_{22} \end{bmatrix} = \begin{bmatrix} 3/2 & -1/2 \\ -1/2 & 5/6 \end{bmatrix} = Z^{-1}$$

11.   The circuit shown below has a z-parameter equivalent circuit but does not have a y-parameter equivalent circuit.

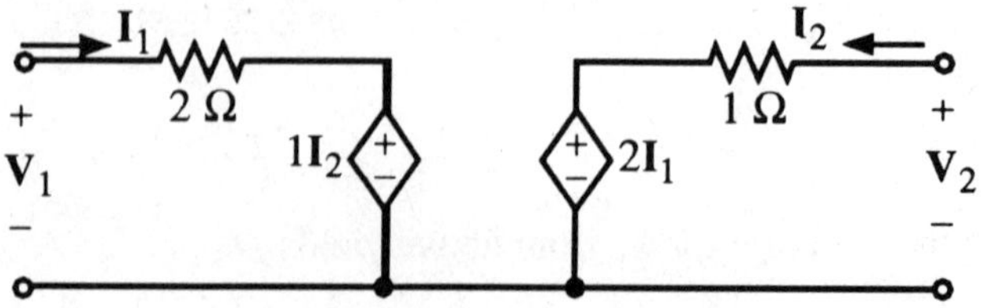

$$Z = \begin{bmatrix} z_{11} & z_{12} \\ z_{21} & z_{22} \end{bmatrix} = \begin{bmatrix} 2 & 1 \\ 2 & 1 \end{bmatrix}$$

Note that $|Z| = 0$ and thus has no y-parameter equivalent circuit.

12. The circuit shown below has a y-parameter equivalent circuit but does not have a z-parameter equivalent circuit.

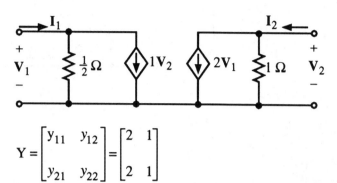

$$Y = \begin{bmatrix} y_{11} & y_{12} \\ y_{21} & y_{22} \end{bmatrix} = \begin{bmatrix} 2 & 1 \\ 2 & 1 \end{bmatrix}$$

Note that $|Y| = 0$ and thus has no z-parameter equivalent circuit.

13. The circuit shown below has neither a y-parameter equivalent circuit nor a z-parameter equivalent circuit.

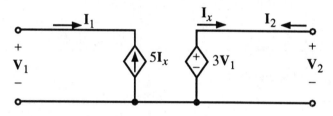

14. Set $h_{11} = 0, V_{1oc} = 0, I_{2sc} = 0, h_{21} = \beta$, and $h_{22} = 0$ to obtain Fig. 16.21

Note that since h_{22} is an admittance, an open circuit is the equivalent of 0 admittance.

15. Find $V_{1o\,csc}$:

Find $I_{2o\,csc}$:

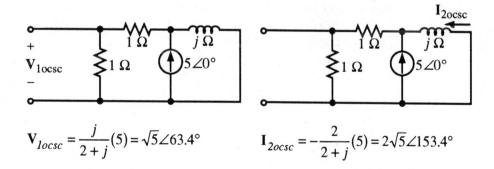

$$V_{1ocsc} = \frac{j}{2+j}(5) = \sqrt{5}\angle 63.4°$$

$$I_{2ocsc} = -\frac{2}{2+j}(5) = 2\sqrt{5}\angle 153.4°$$

Find h_{11} and h_{21} :

Find h_{12} and h_{22} :

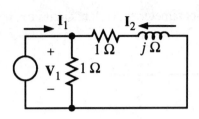

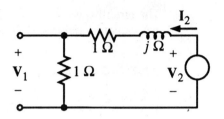

$$V_1 = \left[1 \| (1+j)\right]I_1$$

$$\Rightarrow h_{11} = \frac{3}{5} + j\frac{1}{5}$$

$$I_2 = -\left(\frac{1}{2+j}\right)I_1$$

$$\Rightarrow h_{21} = -\frac{2}{5} + j\frac{1}{5}$$

$$V_1 = \frac{1}{2+j} V_2$$

$$\Rightarrow h_{12} = \frac{2}{5} - j\frac{1}{5}$$

$$V_2 = (2+j)I_2$$

$$\Rightarrow h_{22} = \frac{2}{5} - j\frac{1}{5}$$

$$\Rightarrow \begin{bmatrix} V_1 \\ I_2 \end{bmatrix} = \begin{bmatrix} \frac{3}{5} + j\frac{1}{5} & \frac{2}{5} - j\frac{1}{5} \\ -\frac{2}{5} + j\frac{1}{5} & \frac{2}{5} - j\frac{1}{5} \end{bmatrix} \cdot \begin{bmatrix} I_1 \\ V_2 \end{bmatrix} + \begin{bmatrix} 2.24\angle 63.4° \\ 4.47\angle -26.6° \end{bmatrix}$$

16. Given the circuit shown below:

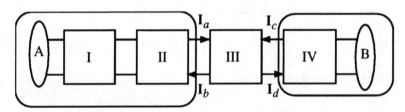

$$I_a + I_b = 0 \ : \ I_c + I_d = 0 \quad \text{Then } I_a + I_b + I_c + I_d = 0.$$

Thus network III satisfies the two-port condition.

17. Given the circuit shown below:

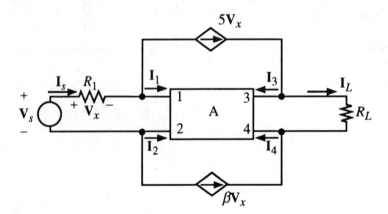

$$\text{KCL}: \quad \mathbf{I}_s = 5\mathbf{V}_x + \mathbf{I}_1 \ \Rightarrow \ \mathbf{I}_1 = \mathbf{I}_s - 5\mathbf{V}_x$$

$$\text{KCL}: \quad \mathbf{I}_s + \mathbf{I}_2 + \beta\mathbf{V}_x = 0 \ \Rightarrow \ \mathbf{I}_2 = -\beta\mathbf{V}_x - \mathbf{I}_s$$

$$\text{KCL}: \quad 5\mathbf{V}_x = \mathbf{I}_L + \mathbf{I}_3 \ \Rightarrow \ \mathbf{I}_3 = 5\mathbf{V}_x - \mathbf{I}_L$$

$$\text{KCL}: \quad \beta\mathbf{V}_x = \mathbf{I}_4 - \mathbf{I}_L \ \Rightarrow \ \mathbf{I}_4 = \beta\mathbf{V}_x + \mathbf{I}_L$$

Note that $\mathbf{I}_1 + \mathbf{I}_2 + \mathbf{I}_3 + \mathbf{I}_4 = 0 \ \Rightarrow \$ Network A satisfies the two-port property.

18.   Given the circuit shown below:

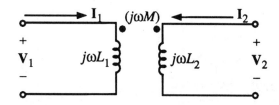

$$\mathbf{V}_1 = j\omega L_1 \mathbf{I}_1 + j\omega M \mathbf{I}_2 \ : \ \mathbf{V}_2 = j\omega M \mathbf{I}_1 + j\omega L_2 \mathbf{I}_2$$

$$\Rightarrow \ Z = \begin{bmatrix} z_{11} & z_{12} \\ z_{21} & z_{22} \end{bmatrix} = \begin{bmatrix} j\omega L_1 & j\omega M \\ j\omega M & j\omega L_2 \end{bmatrix}$$

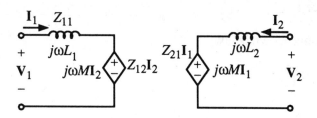

$$\Rightarrow \ Y = \begin{bmatrix} y_{11} & y_{12} \\ y_{21} & y_{22} \end{bmatrix} = \begin{bmatrix} \dfrac{j\omega L_2}{|Z|} & -\dfrac{j\omega M}{|Z|} \\ -\dfrac{j\omega M}{|Z|} & \dfrac{j\omega L_1}{|Z|} \end{bmatrix}$$

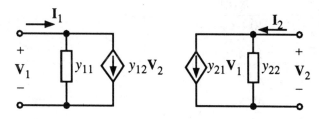

$$H = \begin{bmatrix} h_{11} & h_{12} \\ h_{21} & h_{22} \end{bmatrix} = \begin{bmatrix} \dfrac{|Z|}{j\omega L_2} & \dfrac{M}{L_2} \\ -\dfrac{M}{L_2} & \dfrac{1}{j\omega L_2} \end{bmatrix}$$

19.   Given the circuit shown below:

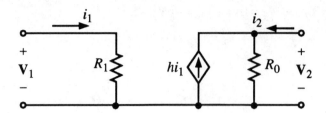

The *h*-parameter circuit is obtained by inspection

$$H = \begin{bmatrix} h_{11} & h_{12} \\ h_{21} & h_{22} \end{bmatrix} = \begin{bmatrix} R_1 & 0 \\ -h & 1/R_o \end{bmatrix} : |H| = \frac{R_1}{R_o}$$

$$\Rightarrow \boxed{\begin{bmatrix} V_1 \\ I_2 \end{bmatrix} = \begin{bmatrix} R_1 & 0 \\ -h & 1/R_o \end{bmatrix} \cdot \begin{bmatrix} I_1 \\ V_2 \end{bmatrix}}$$

$$Z = \begin{bmatrix} z_{11} & z_{12} \\ z_{21} & z_{22} \end{bmatrix} = \begin{bmatrix} R_1 & 0 \\ hR_o & R_o \end{bmatrix} \Rightarrow \boxed{\begin{bmatrix} V_1 \\ V_2 \end{bmatrix} = \begin{bmatrix} R_1 & 0 \\ hR_o & R_o \end{bmatrix} \cdot \begin{bmatrix} I_1 \\ I_2 \end{bmatrix}}$$

$$Y = \begin{bmatrix} y_{11} & y_{12} \\ y_{21} & y_{22} \end{bmatrix} = \begin{bmatrix} 1/R_1 & 0 \\ -h/R_1 & 1/R_o \end{bmatrix} \Rightarrow \boxed{\begin{bmatrix} I_1 \\ I_2 \end{bmatrix} = \begin{bmatrix} 1/R_1 & 0 \\ -h/R_1 & 1/R_o \end{bmatrix} \cdot \begin{bmatrix} V_1 \\ V_2 \end{bmatrix}}$$

20.   Given the circuit shown below:

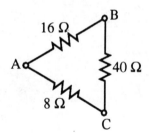

$$R_1 = \frac{R_b R_c}{R_a + R_b + R_c} = \frac{8(16)}{40 + 8 + 16} \Rightarrow R_1 = 2\,\Omega$$

$$R_2 = \frac{R_c R_a}{R_a + R_b + R_c} = \frac{16(40)}{64} \Rightarrow R_2 = 10\,\Omega$$

$$R_3 = \frac{R_a R_b}{R_a + R_b + R_c} = \frac{40(8)}{64} \Rightarrow R_3 = 5\,\Omega$$

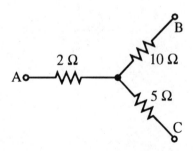

21.   Given the circuit shown below:

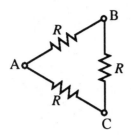

$$R_1 = R_2 = R_3 = \frac{R_b R_c}{R_a + R_b + R_c} = \frac{R^2}{3R} = \frac{R}{3}$$

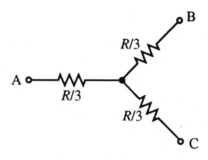

22.   Given the circuit shown below:

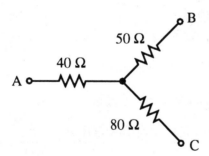

$$R_a = \frac{R_1 R_2 + R_2 R_3 + R_3 R_1}{R_1} = \frac{40(50) + 50(80) + 80(40)}{40} = \frac{9200}{40} = 230\,\Omega$$

$$R_b = \frac{R_1 R_2 + R_2 R_3 + R_3 R_1}{R_2} = \frac{9200}{50} = 184\,\Omega$$

$$R_c = \frac{R_1 R_2 + R_2 R_3 + R_3 R_1}{R_3} = \frac{9200}{80} = 115\,\Omega$$

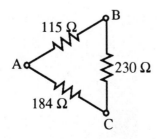

171

23.   Given the circuit shown below:

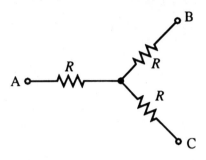

$$R_a = R_b = R_c = \frac{3R^2}{R} = 3R$$

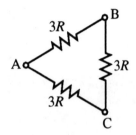

24.   Given the circuit shown below:

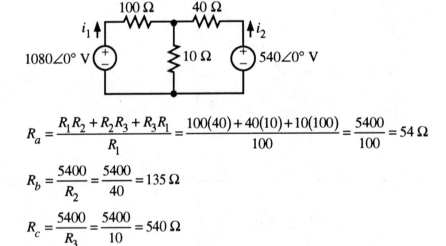

$$R_a = \frac{R_1R_2 + R_2R_3 + R_3R_1}{R_1} = \frac{100(40) + 40(10) + 10(100)}{100} = \frac{5400}{100} = 54\ \Omega$$

$$R_b = \frac{5400}{R_2} = \frac{5400}{40} = 135\ \Omega$$

$$R_c = \frac{5400}{R_3} = \frac{5400}{10} = 540\ \Omega$$

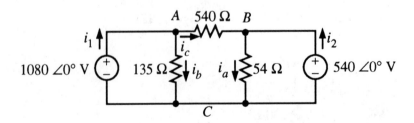

$$v_{ab} = 1080\angle 0° - 540\angle 0° = 540\angle 0° \text{ V}$$

$$i_1 = i_b + i_c = \frac{1080\angle 0°}{135} + \frac{540\angle 0°}{540} \quad \Rightarrow \quad \boxed{i_1 = 9\angle 0° \text{ A}}$$

$$i_2 = i_a - i_c = \frac{540\angle 0°}{54} - \frac{540\angle 0°}{540} \quad \Rightarrow \quad \boxed{i_2 = 9\angle 0° \text{ A}}$$

25.   Given the circuit shown below:

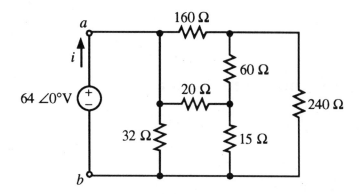

Make an equivalent Δ for the Y composed of R_1, R_2, and R_3:

$$R_a = \frac{20(60) + 60(15) + 15(20)}{20} = \frac{2400}{20} = 120 \ \Omega$$

$$R_b = \frac{2400}{60} = 40 \ \Omega$$

$$R_c = \frac{2400}{15} = 160 \ \Omega$$

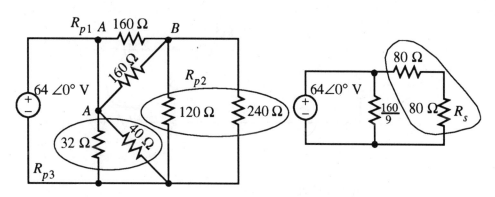

$$R_{p1} = 160 \| 160 = 80 \ \Omega \qquad\qquad R_s = 80 + 80 = 160 \ \Omega$$

$$R_{p2} = 120 \| 240 = 80 \ \Omega \qquad\qquad R_{ab} = 17.777 \| 160 = 16 \ \Omega$$

$$R_{p3} = 32 \| 40 = 17.777 \ \Omega \qquad\qquad \boxed{i = \frac{v_{ab}}{R_{ab}} = \frac{64}{16} = 4 \text{ A}}$$

CHAPTER 17

Mutual Inductance and Transformers

Exercises

1.

$$v_{ab} = L_1 \frac{di_y}{dt} + M \frac{d}{dt}\left[i_y - i_x\right]$$

$$= \left(L_1 + M\right)\frac{di_y}{dt} - M \frac{di_x}{dt}$$

$$v_{cb} = -L_2 \frac{d}{dt}\left[i_y - i_x\right] - M \frac{di_y}{dt}$$

$$= -\left(L_2 + M\right)\frac{di_y}{dt} + L_2 \frac{di_x}{dt}$$

2.

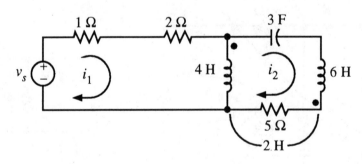

Assign mesh currents as shown.
KVL applied to the first mesh yields

$$-v_s + 3i_1 + 4\frac{d}{dt}\left(i_1 - i_2\right) + 2\frac{d}{dt}\left(-i_2\right) = 0$$

For the second mesh, KVL gives

$$4\frac{d}{dt}\left(i_2 - i_1\right) + 2\frac{d}{dt}\left(i_2\right) + \frac{1}{3}\int_{-\infty}^{t} i_2 d\lambda + 6\frac{d}{dt}i_2 + 2\frac{d}{dt}\left(i_2 - i_1\right) + 5i_2 = 0$$

We can write these two KVL equations as

$$\begin{bmatrix} 4\dfrac{d}{dt}+3 & -6\dfrac{d}{dt} \\[2ex] -6\dfrac{d}{dt} & 14\dfrac{d}{dt}+5+\dfrac{1}{3}\displaystyle\int_{-\infty}^{t}d\lambda \end{bmatrix}\begin{bmatrix} i_1 \\[1ex] i_2 \end{bmatrix}=\begin{bmatrix} v_s \\[1ex] 0 \end{bmatrix}$$

3. $L_1 = 16\text{ H},\ L_2 = 9\text{ H},\ \text{and }M = 5\text{ H}$

$$k = \frac{M}{\sqrt{L_1 L_2}} = \frac{5}{\sqrt{(16)(9)}} = \frac{5}{12} \ \Rightarrow \ \boxed{k = 0.4167}$$

4. $L_1 = 25\text{ H},\ L_2 = 9\text{ H},\ \text{and }k = 0.4$

$$M = k\sqrt{L_1 L_2} = 0.4\sqrt{(25)(9)} \ \Rightarrow \ \boxed{M = 6\text{ H}}$$

5. $i_{ab} = i_y = 4e^{2t}\text{ A}$

KCL gives

$$i_{bc} = i_y - i_x = 4e^{2t} - 2e^{2t} = 2e^{2t}$$

$$w = \frac{1}{2}L_1 i_{ab}^2 + \frac{1}{2}L_2 i_{bc}^2 + M i_{ab} i_{bc}$$

$$= \frac{1}{2}(16)(16e^{4t}) + \frac{1}{2}(9)(4e^{4t}) + 5(4e^{2t})(2e^{2t})$$

$$= 186e^{4t}\text{ J}$$

6. From Exercise 5

$$i_{ab} = 4e^{2t}\text{ A}, \quad i_{bc} = 2e^{2t}\text{ A}$$

$$w = \frac{1}{2}L_1 i_{ab}^2 + \frac{1}{2}L_2 i_{bc}^2 - M i_{ab} i_{bc}$$

$$= \frac{1}{2}(16)(16e^{4t}) + \frac{1}{2}(9)(4e^{4t}) - 5(4e^{2t})(2e^{2t})$$

$$= 106e^{4t}\text{ J}$$

7.

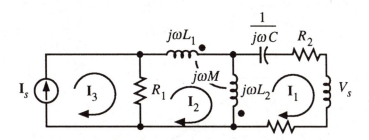

Use the short cut procedure.

KVL applied to mesh 1 gives

$$\mathbf{V}_s + j\omega L_2(\mathbf{I}_1 - \mathbf{I}_2) + j\omega M(-\mathbf{I}_2) + \frac{1}{j\omega C}\mathbf{I}_1 + R_2\mathbf{I}_1 = 0$$

For mesh 2, KVL gives

$$R_1(\mathbf{I}_2 - \mathbf{I}_s) + j\omega L_1(\mathbf{I}_2) + j\omega M(\mathbf{I}_2 - \mathbf{I}_1) + j\omega L_2(\mathbf{I}_2 - \mathbf{I}_1) + j\omega M(\mathbf{I}_2) = 0$$

We arrange these two KVL equations as

$$\begin{bmatrix} j\omega L_2 + R_2 + \frac{1}{j\omega C} & -j\omega(L_2 + M) \\ -j\omega(L_2 + M) & j\omega(L_1 + L_2 + 2M) + R_1 \end{bmatrix}\begin{bmatrix} \mathbf{I}_1 \\ \mathbf{I}_2 \end{bmatrix} = \begin{bmatrix} -\mathbf{V}_s \\ R_1\mathbf{I}_s \end{bmatrix}$$

8.

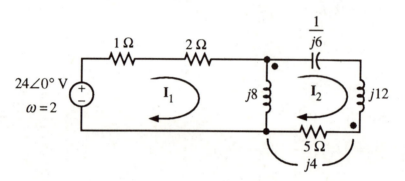

For mesh 1 KVL gives

$$-24 + 1\mathbf{I}_1 + 2\mathbf{I}_1 + j8(\mathbf{I}_1 - \mathbf{I}_2) + j4(-\mathbf{I}_2) = 0$$

For mesh 2 KVL gives

$$j8(\mathbf{I}_2 - \mathbf{I}_1) + j4(\mathbf{I}_2) + \frac{1}{j6}\mathbf{I}_2 + j12\mathbf{I}_2 + j4(\mathbf{I}_2 - \mathbf{I}_1) + 5\mathbf{I}_2 = 0$$

We can write these two equations as a single matrix equation

$$\begin{bmatrix} 3 + j8 & -j12 \\ -j12 & 5 + j\frac{167}{6} \end{bmatrix}\begin{bmatrix} \mathbf{I}_1 \\ \mathbf{I}_2 \end{bmatrix} = \begin{bmatrix} 24\angle 0° \\ 0 \end{bmatrix}$$

9. $\mathbf{Z}_{11} = R_1 + j\omega L_1 \cong j\omega L_1$

$\mathbf{Z}_{22} = R_2 + j\omega L_2 \cong j\omega L_2$

$\mathbf{Z}_{12} = \mathbf{Z}_{21} = j\omega M$

(a) $\mathbf{Z}_1 = \mathbf{Z}_{11} - \mathbf{Z}_{12}\mathbf{Z}_{21}\dfrac{1}{\mathbf{Z}_{22} + \mathbf{Z}_L} = j\omega L_1 - (j\omega M)^2\dfrac{1}{j\omega L_2 + \mathbf{Z}_L} = j\omega L_1 + \dfrac{\omega^2 M^2}{j\omega L_2 + \mathbf{Z}_L}$

(b) $M = k\sqrt{L_1 L_2}$; $\mathbf{Z}_1 = j\omega L_1 + \dfrac{\omega^2 k^2 L_1 L_2}{j\omega L_2 + \mathbf{Z}_L}$

(c) $L_2 = (n_2/n_1)^2 L_1$

$$\mathbf{Z}_1 = j\omega L_1 + \frac{\omega^2 k^2 (n_2/n_1)^2 L_1^2}{j\omega(n_2/n_1)^2 L_1^2 + \mathbf{Z}_L}$$

10. $\mathbf{Z}_L = R_L$

(a) $\mathbf{Z}_1 = j\omega L_1 + \dfrac{\omega^2 M^2}{R_L + j\omega L_2}$

(b) $\mathbf{H}_v(j\omega) = \dfrac{\mathbf{V}_2}{\mathbf{V}_1} = \dfrac{\mathbf{Z}_{21}\mathbf{Z}_L}{\mathbf{Z}_{11}(\mathbf{Z}_{22} + \mathbf{Z}_L) - \mathbf{Z}_{12}\mathbf{Z}_{21}} = \dfrac{j\omega M R_L}{j\omega L_1(j\omega L_2 + R_L) + \omega^2 M^2}$

(c) $\mathbf{H}_I(j\omega) = \dfrac{\mathbf{I}_2}{\mathbf{I}_1} = -\dfrac{\mathbf{Z}_{21}}{\mathbf{Z}_{22} + \mathbf{Z}_L} = -\dfrac{j\omega M}{j\omega L_2 + R_L}$

(d) $\mathbf{G}_I(j\omega) = -\dfrac{\mathbf{V}_2 \mathbf{I}_2^*}{\mathbf{V}_1 \mathbf{I}_1^*} = -\dfrac{j\omega M R_L}{j\omega L_1(j\omega L_2 + R_L) + \omega^2 M^2} \cdot \dfrac{j\omega M}{R_L - j\omega L_2}$

$$= \dfrac{\omega^2 M^2 R_L}{\left[j\omega L_1(R_L + j\omega L_2) + \omega^2 M^2\right](R_L - j\omega L_2)}$$

11. $L_a = L_1 - M = 1\ \text{H} : L_b = L_2 - M = 39\ \text{H} : L_c = M = 1\ \text{H}$

12. $L_a = L_1 - M = 0\ \text{(a short circuit)} : L_b = L_2 - M = 38\ \text{H} : L_c = M = 2\ \text{H}$

13. $L_a = L_1 - M = -1\ \text{H} : L_b = L_2 - M = 37\ \text{H} : L_c = M = 3\ \text{H}$

   $L_a < 0 \Rightarrow$ An equivalent T model exists only for a fixed frequency.

14. $L_a = L_1 - M = -4\ \text{H} : L_b = L_2 - M = 34\ \text{H} : L_c = M = 6\ \text{H}$

   $L_a < 0 \Rightarrow$ An equivalent T model exists only for a fixed frequency.

Refer to the circuits shown below for Exercises 15 - 18.

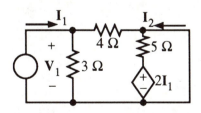
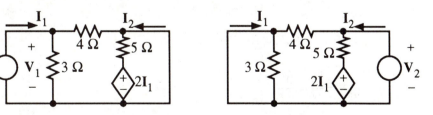

15. $L_A = \dfrac{L_1 L_2 - M^2}{L_2 - M} = \dfrac{79}{39}\ \text{H} = 2.026\ \text{H} : L_B = \dfrac{L_1 L_2 - M^2}{L_1 - M} = 79\ \text{H}$

$$L_C = \frac{L_1 L_2 - M^2}{M} = 79 \text{ H}$$

16.
$$L_A = \frac{L_1 L_2 - M^2}{L_2 - M} = \frac{76}{38} \text{ H} = 2 \text{ H} \;:\; L_B = \frac{L_1 L_2 - M^2}{L_1 - M} \rightarrow \infty \text{ (open circuit)}$$

$$L_C = \frac{L_1 L_2 - M^2}{M} = 38 \text{ H}$$

17.
$$L_A = \frac{L_1 L_2 - M^2}{L_2 - M} = \frac{71}{37} \text{ H} = 1.919 \text{ H} \;:\; L_B = \frac{L_1 L_2 - M^2}{L_1 - M} = -71 \text{ H}$$

$$L_C = \frac{L_1 L_2 - M^2}{M} = 23.67 \text{ H}$$

$L_B < 0 \;\Rightarrow\;$ A II equivalent circuit exists only for a fixed frequency.

18.
$$L_A = \frac{L_1 L_2 - M^2}{L_2 - M} = \frac{44}{34} \text{ H} = 1.294 \text{ H} \;:\; L_B = \frac{L_1 L_2 - M^2}{L_1 - M} = -22 \text{ H}$$

$$L_C = \frac{L_1 L_2 - M^2}{M} = 7.333 \text{ H}$$

$L_B < 0 \;\Rightarrow\;$ A II equivalent circuit exists only for a fixed frequency.

19.
$$\mathbf{S} = \mathbf{S}_1 + \mathbf{S}_2 = \mathbf{V}_1 \mathbf{I}_1^* + \mathbf{V}_2 \mathbf{I}_2^*$$

$$= \mathbf{V}_1 \mathbf{I}_1^* + \frac{n_2}{n_1} \mathbf{V}_1 \left(-\frac{n_1}{n_2} \mathbf{I}_1 \right)^*$$

$$= \mathbf{V}_1 \mathbf{I}_1^* - \mathbf{V}_1 \mathbf{I}_1^* = 0$$

where n_1 / n_2 is real.

20.   (a)
$$\mathbf{S} = \mathbf{S}_a + \mathbf{S}_b + \mathbf{S}_c$$

$$= (2 + j1) + (3 + j1) + (5 + j0)$$

$$= (10 + j2) KVA = 10.1980\angle 11.3099° KVA$$

(b)
$$I_{rms} = \frac{|\mathbf{S}|}{V} = \frac{10.1980 \times 10^3}{13.8 \times 10^3} = 0.739 \text{ A}$$

(c)
$$I_{rms} = \frac{|\mathbf{S}|}{V} = \frac{10.1980 \times 10^3}{240} = 42.5 \text{ A}$$

21.   Given the circuit shown below:

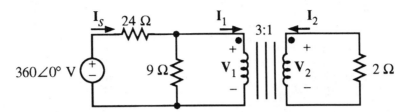

(a)   Reflect the secondary impedances into the primary:

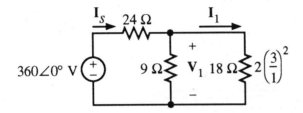

Then $I_s = \dfrac{360\angle 0°}{30} = 12\angle 0°$ A

(b) $V_1 = 6I_s = 72\angle 0°$ V

(c) $I_1 = \dfrac{1}{18}V_1 = 4\angle 0°$ A

(d) $V_2 = \dfrac{1}{3}V_1 = 24\angle 0°$ V

(e) $I_2 = -3I_1 = 12\angle 180°$ A

(f) $Z_{in} = R_s = 30\ \Omega$

22.   Refer to the circuit shown in Exercise 21.

Disconnect the 2-Ω resistor. Then voltage divider gives the following:

$$V_1 = \frac{9}{24+9}(360\angle 0°) = \frac{1080}{11}\angle 0°\ \text{V} \Rightarrow \boxed{V_{TH} = V_2 = \frac{1}{3}\left(\frac{1080}{11}\right) = 32.73\angle 0°\ \text{V}}$$

Note that V_{TH} is calculated with $I_2 = 0$.

Now, with the voltage source set to zero, the 9-Ω and 24-Ω resistors can be combined in parallel to yield

$$R_p = 24 \| 9 = \frac{72}{11}\ \Omega$$

Then the ideal transformer gives

$$R_{Th} = \left(\frac{1}{3}\right)^2 R_p = 0.723\,\Omega$$

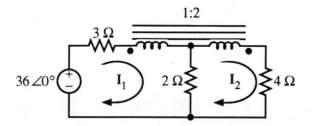

A quick check with the 2-Ω resistor connected yields $I_2 = -12$ A. This result agrees with Exercise 21.

23.   Given the circuit shown below:

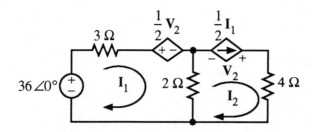

Replace this circuit with the model of Figure 17.11

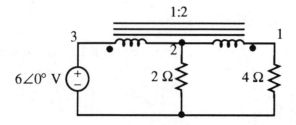

KVL: $36 = 3I_1 + \dfrac{1}{2}\left[4\left(\dfrac{1}{2}I_1\right) + 2\left(\dfrac{1}{2}I_1 - I_1\right)\right] + 2\left(I_1 - \dfrac{1}{2}I_1\right)$

$\Rightarrow$ $\boxed{I_1 = 8\angle 0^\circ \text{ A}}$ $\Rightarrow$ $\boxed{I_2 = \dfrac{1}{2}I_1 = 4\angle 0^\circ \text{ A}}$

24.   Given the circuit shown below:

Using the transformer model of Fig. 17.11, the circuit becomes

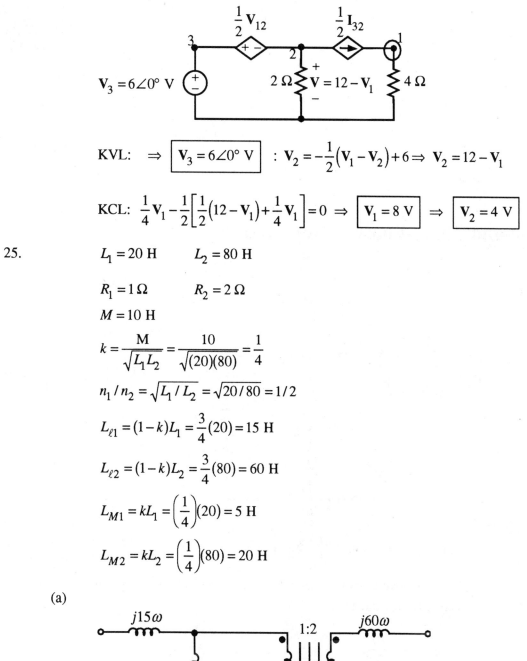

$$\frac{1}{2}\mathbf{V}_{12} \qquad \frac{1}{2}\mathbf{I}_{32}$$

$$\mathbf{V}_3 = 6\angle 0° \text{ V} \qquad 2\ \Omega \quad \mathbf{V} = 12 - \mathbf{V}_1 \qquad 4\ \Omega$$

KVL: $\Rightarrow \boxed{\mathbf{V}_3 = 6\angle 0° \text{ V}}$: $\mathbf{V}_2 = -\frac{1}{2}(\mathbf{V}_1 - \mathbf{V}_2) + 6 \Rightarrow \mathbf{V}_2 = 12 - \mathbf{V}_1$

KCL: $\frac{1}{4}\mathbf{V}_1 - \frac{1}{2}\left[\frac{1}{2}(12 - \mathbf{V}_1) + \frac{1}{4}\mathbf{V}_1\right] = 0 \Rightarrow \boxed{\mathbf{V}_1 = 8 \text{ V}} \Rightarrow \boxed{\mathbf{V}_2 = 4 \text{ V}}$

25.

$$L_1 = 20 \text{ H} \qquad L_2 = 80 \text{ H}$$

$$R_1 = 1\ \Omega \qquad R_2 = 2\ \Omega$$

$$M = 10 \text{ H}$$

$$k = \frac{M}{\sqrt{L_1 L_2}} = \frac{10}{\sqrt{(20)(80)}} = \frac{1}{4}$$

$$n_1 / n_2 = \sqrt{L_1 / L_2} = \sqrt{20/80} = 1/2$$

$$L_{\ell 1} = (1 - k)L_1 = \frac{3}{4}(20) = 15 \text{ H}$$

$$L_{\ell 2} = (1 - k)L_2 = \frac{3}{4}(80) = 60 \text{ H}$$

$$L_{M1} = kL_1 = \left(\frac{1}{4}\right)(20) = 5 \text{ H}$$

$$L_{M2} = kL_2 = \left(\frac{1}{4}\right)(80) = 20 \text{ H}$$

(a)

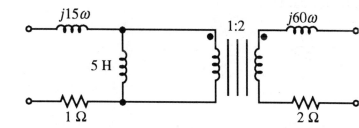

(b)

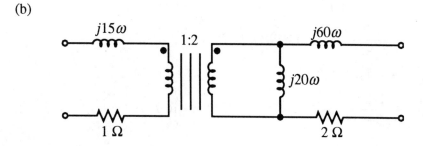

CHAPTER 18

Single-and Three-Phase Power Circuits

Exercises

1.

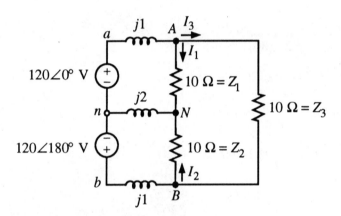

Because $\mathbf{V}_{an} = \mathbf{V}_{nb}$ and $\mathbf{Z}_1 = \mathbf{Z}_2$, $\mathbf{I}_{nN} = 0$.

We can remove the neutral wire $Z_n = j2$ with no effect on the current. There is an effective resistance of

$$R = \frac{1}{\frac{1}{20} + \frac{1}{10}} = \frac{20}{3} \, \Omega$$

connected between A and B.

$$\mathbf{I}_{aA} = \frac{120\angle 0° - 120\angle 180°}{R + j2} = \frac{240}{6.667 + j2} = \frac{240}{6.9602\angle 16.7°} = 34.48\angle -16.7° \text{ A}$$

$$\mathbf{I}_1 = \frac{1/20}{1/20 + 1/10} \mathbf{I}_{aA} = 11.49\angle -16.7° \text{ A}$$

$$\mathbf{I}_2 = -\mathbf{I}_1 = 11.49\angle 163.3° \text{ A}$$

$$\mathbf{I}_3 = \frac{1/20}{1/20 + 1/10} \, 34.49\angle -16.7° = 22.99\angle -16.7° \text{ A}$$

$$\mathbf{V}_{AN} = 10\mathbf{I}_1 = 114.9\angle -16.7° \text{ V}$$

$$\mathbf{V}_{BN} = 10\mathbf{I}_2 = 114.9\angle 163.3° \text{ V}$$

$$\mathbf{V}_{Nn} = j2(0) = 0 \text{ V}$$

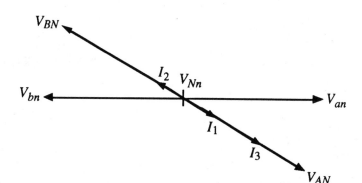

2. $$v_{aa'} = 1600\cos(2\pi 60 t + 45°)$$

$$\mathbf{V}_{aa'} = \frac{1600}{\sqrt{2}}\angle -45° = 1131\angle 45° \text{ V}$$

$$\mathbf{V}_{bb'} = \mathbf{V}_{aa'}(1\angle -120°) = (1131\angle 45°)(1\angle -120°) = (1131\angle -75° \text{ V})$$

$$\mathbf{V}_{cc'} = \mathbf{V}_{aa'}(1\angle -240°) = (1131\angle 45°)(1\angle -240°) = (1131\angle -195° \text{ V})$$

3. $$\mathbf{V}_{aa'} = 240\angle 30° \text{ V is given}$$

(a)   For a positive phase sequence

$$\mathbf{V}_{aa'} = 240\angle 30° \text{ V}$$

$$\mathbf{V}_{bb'} = 240\angle 30° - 120° = 240\angle -90° \text{ V}$$

$$\mathbf{V}_{cc'} = 240\angle 30° - 240° = 240\angle -210° \text{ V}$$

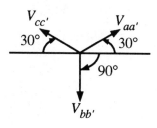

(b)   For a negative phase sequence

$$\mathbf{V}_{aa'} = 240\angle 30° \text{ V}$$

$$\mathbf{V}_{bb'} = 240\angle 30° + 120° = 240\angle 150° \text{ V}$$

$$\mathbf{V}_{cc'} = 240\angle 30° + 240° = 240\angle 270° \text{ V}$$

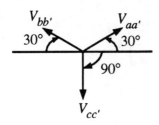

4.   Three-phase systems are specified in terms of the rms value of the line voltages.

(a) $V_L = 480$ V

(b) $V_{LN} = \dfrac{1}{\sqrt{3}}480 = 277$ V

(c) $V_{peak} = \sqrt{2}V_L = \sqrt{2}\,480 = 679$ V

(d)   For a positive phase sequence, the line voltage leads the line-to-neutral voltage by 30°.

5. $\mathbf{V}_{BC} = (1\angle -120°)\mathbf{V}_{AB} = 208\angle -120°$ V

$\mathbf{V}_{CA} = (1\angle 120°)\mathbf{V}_{AB} = 208\angle 120°$ V or $208\angle -240°$ V

6. $\mathbf{V}_{AN} = \left(\dfrac{1}{\sqrt{3}\angle 30°}\right)\mathbf{V}_{AB} = 120\angle -30°$ V : $\mathbf{V}_{BN} = 120\angle -150°$ V

$\mathbf{V}_{CN} = 120\angle -270°$ or $\mathbf{V}_{CN} = 120\angle 90°$ V

7. $\mathbf{I}_{AN} = \dfrac{\mathbf{V}_{AN}}{\mathbf{Z}_{LN}} = \dfrac{120\angle -30°}{20\angle 15°} = 6\angle -45°$ A : $\mathbf{I}_{BN} = 6\angle -165°$ A : $\mathbf{I}_{CN} = 6\angle 75°$ A

8. $\mathbf{S} = 3\mathbf{S}_\phi = 3\mathbf{V}_{AN}\mathbf{I}_{AN}^{\;*} = 3(120\angle -30°)(6\angle -45°)^* = 2160\angle 15°$ VA

9. $V_L = \sqrt{3}(120) = 207.8 \approx 208$ V

10. $\mathbf{I}_{AB} = \dfrac{\mathbf{V}_{AB}}{\mathbf{Z}_{LL}} = \dfrac{240\angle 0°}{20\angle 15°} = 12\angle -15°$ A : $\mathbf{I}_{BC} = 12\angle -135°$ A : $\mathbf{I}_{CA} = 12\angle 105°$ A

11. $\mathbf{I}_{aA} = \mathbf{I}_{AB}\left(\sqrt{3}\angle -30°\right) = 20.78\angle -45°$ A

: $\mathbf{I}_{bB} = 20.78\angle -165°$ A : $\mathbf{I}_{cC} = 20.78\angle 75°$ A

12. $\mathbf{S} = 3\dfrac{V_L^2}{\mathbf{Z}_{LL}^{\;*}} = 3\dfrac{240^2}{(20\angle 15°)^*} = 8640\angle 15°$ VA

13.   (a) $\mathbf{I}_{AN} = \mathbf{I}_{aA} = 10\angle -90°$ A

$$V_{AN} = V_{AB}\left(\frac{1}{\sqrt{3}}\angle -30°\right) = (200\angle 0°)\left(\frac{1}{\sqrt{3}}\angle -30°\right) = 115.47\angle -30° \text{ V}$$

$$S = 3V_{AN}I_{AN}^*$$

$$= 3(115.47\angle -30°)(10\angle 90°) = 3464\angle 60° \text{ VA} = (1732 + j3000) \text{ VA}$$

(b) $P = \mathcal{R}e\{S\} = 1732 \text{ W}$

14.   (a) $S_\Delta = 3\dfrac{|V_L|^2}{Z_\Delta^*} = 3\dfrac{480^2}{3+j4} = 138.24\angle -53.13° \text{ kVA} = (82.944 - j110.592) \text{ kVA}$

(b) $S_Y = \dfrac{|V_L|^2}{Z_Y^*} = \dfrac{480^2}{3-j4} = 46.08\angle 53.13° \text{ kVA} = (27.648 + j36.864) \text{ kVA}$

(c) $S = S_\Delta + S_Y = (110.592 - j73.728) = 132.92\angle -33.69° \text{ kVA}$

(d) $I_L = \dfrac{|S|}{\sqrt{3}\,V_L} = \dfrac{138.24\times 10^3}{480\sqrt{3}} = 166.28 \text{ A}$

(e) $I_L = \dfrac{|S|}{\sqrt{3}\,V_L} = \dfrac{46.08\times 10^3}{480\sqrt{3}} = 55.43 \text{ A}$

(f) $I_L = \dfrac{|S|}{\sqrt{3}\,V_L} = \dfrac{132.92\times 10^3}{480\sqrt{3}} = 159.88 \text{ A}$

15. $PF = 0.8$ lagging is given

$$\theta = \text{arc}\cos(PF) = 36.87°$$
$$|S| = 25 \text{ kVA is given}$$

(a) $S_L = 25\angle 36.87° \text{ kVA} = (20 + j15) \text{ kVA}$

(b) $I_L = \dfrac{|S|}{\sqrt{3}\,V_L} = \dfrac{25000}{480\sqrt{3}} = 30.07 \text{ A}$

(c) $S_{new} = S_L + S_C = (20 + j15 + jQ) \text{ kVA}$

$$\theta = \text{arc}\cos(0.9) = 25.84°$$
$$\frac{15 + Q}{20} = \tan 25.84° = 0.4843$$
$$Q_c = -5.3136 \text{ kVAR}$$

(d) $\dfrac{15+Q}{20} = -\tan 25.84° = -0.4843$

$Q_d = -24.686$ kVAR

(e)

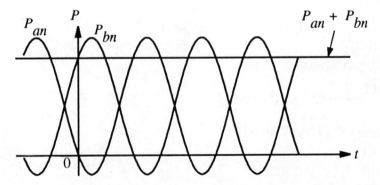

Corrected to 0.9PF leading

16. $p_{an} = v_{an}i_{an} = V_m \cos(\omega t)\left[\dfrac{1}{|\mathbf{Z}|}V_m \cos(\omega t - \theta_z)\right]$

$\Rightarrow$ $\boxed{p_{an} = \dfrac{1}{2|\mathbf{Z}|}V_m^2\left[\cos\theta_z + \cos(2\omega t - \theta_z)\right]}$

$p_{bn} = v_{bn}i_{bn} = V_m \cos(\omega t - 90°)\left[\dfrac{1}{|\mathbf{Z}|}V_m \cos(\omega t - 90° - \theta_z)\right]$

$\Rightarrow$ $\boxed{p_{bn} = \dfrac{1}{2|\mathbf{Z}|}V_m^2\left[\cos(\theta_z) - \cos(2\omega t - \theta_z)\right]}$

$\Rightarrow$ $\boxed{p = p_{an} + p_{bn} = \dfrac{1}{|\mathbf{Z}|}V_m^2 \cos(\theta_z) = P_{av}}$

A plot is shown below:

17. (a) $P_A = \mathcal{R}e\{\mathbf{V}_{AB}\mathbf{I}_{aA}^*\} = 3.2 \text{ kW}$

$P_C = \mathcal{R}e\{\mathbf{V}_{CB}\mathbf{I}_{cC}^*\} = 1.25 \text{ kW}$

$P = P_A + P_C = 3.2 + 1.25 = 4.45 \text{ kW}$

(b) $\theta = \arctan\left(\sqrt{3}\dfrac{P_C - P_A}{P_C + P_A}\right)$

$= \arctan\left(\sqrt{3}\dfrac{1.25 - 3.2}{1.25 + 3.2}\right) = -37.2°$

$PF = \cos(-37.2°) = 0.797 \text{ leading}$

18. $\mathbf{V}_{AB} = 480\angle 30° \text{ V assumed}$

$\mathbf{I}_{aA} = \dfrac{480/\sqrt{3}}{10\angle 36.87°} = 27.71\angle -36.87°$

$P_A = \mathcal{R}e\{\mathbf{V}_{AB}\mathbf{I}_{aA}^*\} = \mathcal{R}e\{(480\angle 30°)(27.71\angle 36.87°)\}$

$= \mathcal{R}e\{13,300\angle 66.87°\} = 5.22 \text{ kW}$

$\mathbf{V}_{CB} = -\mathbf{V}_{BC} = -(\mathbf{V}_{AB})(1\angle -120°)$

$= -480\angle 30° - 120° = 480\angle 30° - 120° + 180°$

$= 480\angle 90°$

$\mathbf{V}_{cn} = (1/\sqrt{3})480\angle 30° - 240° - 30° = \dfrac{480}{\sqrt{3}}\angle 120°$

$\mathbf{I}_{cC} = \dfrac{(480\angle 120°)/\sqrt{3}}{10\angle 36.87°} = 27.71\angle 83.13°$

$P_c = \mathcal{R}e\{\mathbf{V}_{CB}\mathbf{I}_{cC}^*\} = \mathcal{R}e\{(480\angle 90°)(27.71\angle -83.13°)\}$

$= \mathcal{R}e\{13,300\angle 6.87°\} = 13.20 \text{ kW}$

$P = P_A + P_C = 5.22 + 13.20 = 18.43 \text{ kW}$

$\theta = \arctan\left(\sqrt{3}\dfrac{P_C - P_A}{P_C + P_A}\right)$

$= \arctan\left(\sqrt{3}\dfrac{13.2 - 5.22}{13.2 + 5.22}\right) = 36.87°$

$PF = \cos(36.87°) = 0.8 \text{ lagging}$